UltraUnification and the Generation of the Cosmos

HyperCosmos and Second Kind HyperCosmos
Based on One Primordial Dimension (Element)

Stephen Blaha Ph. D.
Blaha Research

Pingree-Hill Publishing

MMXXIII

ISBN: 979-8-9858266-6-1

Rev. 00/00/01 January 23, 2023

To Margaret

Some Other Books by Stephen Blaha

All the Megaverse! Starships Exploring the Endless Universes of the Cosmos using the Baryonic Force (Blaha Research, Auburn, NH, 2014)

SuperCivilizations: Civilizations as Superorganisms (McMann-Fisher Publishing, Auburn, NH, 2010)

All the Universe! Faster Than Light Tachyon Quark Starships & Particle Accelerators with the LHC as a Prototype Starship Drive Scientific Edition (Pingree-Hill Publishing, Auburn, NH, 2011).

Unification of God Theory and Unified SuperStandard Model THIRD EDITION (Pingree Hill Publishing, Auburn, NH, 2018).

The Exact QED Calculation of the Fine Structure Constant Implies ALL 4D Universes have the Same Physics/Life Prospects (Pingree Hill Publishing, Auburn, NH, 2019).

Integration of General Relativity and Quantum Theory: Octonion Cosmology, GiFT, Creation/Annihilation Spaces CASe, Reduction of Spaces to a Few Fermions and Symmetries in Fundamental Frames (Pingree Hill Publishing, Auburn, NH, 2021).

Passing Through Nature to Eternity ProtoCosmos, HyperCosmos, Unified SuperStandard Theory (Pingree Hill Publishing, Auburn, NH, 2022).

HyperCosmos Fractionation and Fundamental Reference Frame Based Unification: Particle Inner Space Basis of Parton and Dual Resonance Models (Pingree Hill Publishing, Auburn, NH, 2022).

A New UniDimension ProtoCosmos and SuperString F-Theory Relation to the HyperCosmos (Pingree Hill Publishing, Auburn, NH, 2022).

The Cosmic Panorama: ProtoCosmos, HyperCosmos, Unified SuperStandard Theory (UST) Derivation (Pingree Hill Publishing, Auburn, NH, 2022).

Ultimate Origin: ProtoCosmos and HyperCosmos (Pingree Hill Publishing, Auburn, NH, 2022).

Available on Amazon.com, bn.com, Amazon.co.uk and other international web sites as well as at better bookstores.

CONTENTS

FIGURES and TABLES

Introduction

The meaning of Dimension is explored as the book begins with a discussion of the role of Dimension as a primordial element. The book exemplifies the continuing critical role of dimension in the HyperCosmos, in the Second Kind HyperCosmos (introduced here) as well as the other spaces of the Cosmos: the 42 dimension Full HyperUnification space and the 88 space-time dimension UltraUnification (UU) space. There are four levels in the Cosmos.

The new spaces bring the total number of Cosmos spaces to 42 spaces. The Full HyperUnification space, which contains the 20 HyperUnification and Second Kind HyperUnification spaces, has 42 space-time dimensions.

The appearance of 42 and 88 dimensions in the Cosmos evokes conjectures of Dirac and Eddington who noted that Physical constants and properties can be combined to form a number of dimensionless constants with the approximate values of 40 or 80. The 42 dimensions may reflect dimensionless constants of value 40 generated from features of the Full HyperUnification space. The 88 dimensions of UU may reflect dimensionless constants of value 80 generated from features of UU.

The entire Cosmos Theory can be generated from one primordial Dimension in the UU space. The author shows that the one dimension can also be used to define a UniDimension ProtoCosmos Model that generates the spaces of the HyperCosmos.

The one primordial dimension in UU can be fractionated into an infinite dust of fractionated dimensions, which can be shaped to form the one dimension space of the ProtoCosmos.

Dimension shapes the 42 spaces. Dimension leads to the Unified SuperStandard Theory and the Standard Model of Elementary Particles.

As a result the One Primordial Dimension in the UU and the ProtoCosmos is the decisive key to the Cosmos. This book presents a complete description of the Cosmos within the framework described above. Cosmos Theory is conceptual. Reality enters when individual universes are created (instantiated) of one space type or another. Created universes are structured according to Cosmos Theory.

Earlier books such as *The Cosmic Panorama, HyperCosmos Fractionation ...,* and *Passing Through Nature to Eternity* are recommended to the reader for a better understanding of the overall theory.

The history of the search for a fundamental substance is now:

Fire-Earth-Air-Water – Atoms – Quarks, Leptons, … – Dimension

This book covers many related topics as well.

1. Meaning of Dimension

The concept of a primordial dimension that was developed in Blaha (2022g) requires some explanation. We can begin by asserting:

> Dimensions are primitives. They may be counted. But they have no numeric value. They are like the primitives of Euclidean geometry such as point and line. We suggest they are at the deepest level and should be treated as undefined primitives. Everything may be reduced to one primary dimension creating a unified set of spaces and universes. These discoveries are not likely to be the result of a yet deeper level of Physics.

as was done in Blaha (2022g). But the question of the origin of the one dimension based form of the HyperCosmos requires explanation. Also the role of dimension in elementary particle theory (the Unified SuperStandard Theory – UST) must be considered. Chapter 2 begins the consideration of these questions.

1.1 Space-time Dimensions

Spatial dimensions were first seriously considered by Euclid. He defined spatial dimensions by generating them by three perpendicular line segments. More recently Einstein introduced the time dimension through the Special Theory of Relativity and extended it with the General Theory of Relativity. Thus our universe enjoys four space-time dimensions whose meaning is based on Geometry and Relativity.

1.2 Other Dimensions

The possibility of other dimensions has been considered by numbers of authors. Most make either ad hoc assumptions based on symmetries or based on the salutary properties of additional dimensions. Some ascribe other dimensions to spiritual or ethereal causes.

1.3 HyperCosmos Dimensions

HyperCosmos Theory defines spaces with an array of dimensions. The dimensions in the dimension array are divided into a set of space-time dimensions and a set of Internal Symmetry dimensions. *One counts the number of dimensions in each set calling the total the number of dimensions.*

Internal Symmetry dimensions are subdivided into sets of dimensions for irreducible representations of the various Internal Symmetry groups.

Implicit in the definition of dimension arrays is the belief that there is no fundamental difference in the nature of space-time dimensions based on Geometry and Internal Symmetry dimensions based on Group Theory. They differ in their applications in Physics and the "real" world. The many other attempts to create unified theories implicitly assume this view.

All dimension definitions agree with the statement:

Dimensions are primitives. They may be counted. But they have no numeric value.

The dimension arrays of the HyperCosmos assume the unity of all dimensions. The development of HyperUnification in Blaha (2022f) and (2022g) proves the unification of space-time and Internal Symmetry in HyperUnification spaces under (solely) General Relativistic transformations in HyperUnification spaces.

1.4 Role of Dimensions in Physics

We defined dimensions in themselves above. The role of non-negative dimensions in all cases is to define coordinate systems for various purposes. The number of dimensions in a system specifies the number of its coordinates. The only exceptions are systems with a negative number of dimensions. The definition of the role of dimensions can be extended to negative dimensions where the negative value of the total number of dimensions specifies the constraints on the coordinates of a system.

1.4.1 Negative Space-time Dimensions

The concept of negative space-time dimensions in elementary particle physics is new. We considered them in chapter 13 of Blaha (2022c).We defined them in terms of a conventional operational definition of dimension. An understanding of negative dimensions enabled the negative space-time dimensions of the FULL HyperCosmos spectrum to be understood.

The dimension of a space-time is the number of independent parameters needed to define a point. An over-determined space-time has a negative dimension.

In particular, we regard a negative dimension as the equivalent of an over-specified point. Some negative dimension examples are:

r = -3

x = 5 x = 2*1.5 x = 6/2 x = 12/4 thrice over-determined

r = -2

x = 3 x = 2*1.5 x = 6/2 doubly over-determined

r = -1

x = 3 x = 2*1.5 singly over-determined

r = 0

x = 3 determined

r = 1 One dimension space with one free parameter

x free parameter

r = 2 Two dimension space with two free parameters

x, y free parameters

where r is the dimension.

1.4.2 Other Definitions of Dimension

Other definitions are possible. There are many possibilities. We believe the definitions above are best suited for the HyperCosmos and Physics in general.

1.5 Space Dimension Arrays and Universes

The HyperCosmos specifies a dimension array for each of the ten spaces that support the definition of universes. We can define a universe wave function for a space as we did in our earlier books.[1] We can then define universe states |U> for universes. The universe states have space-times and Internal Symmetries defined by the dimension arrays. Particles and interactions are thereby incorporated within a universe state. See Blaha (2022c) and (2022g) as well as our earlier books for more detailed discussion.

The possibility of inter-universe interactions raises many potential situations.

1. A given space may have universes with interactions that enable sub-universes to be created.

2. A space may have many universe states. These universe states may interact or may not interact. If they interact then the possibility of parallel Multiverses arises together with the possibility of inter-universe transfers.

3. Sub-universes may interact within a parent universe.

Thus universes may be viewed as a type of particle—possibly with interactions—and undergoing Hubble-like expansion after creation. This viewpoint was explored in detail in Blaha (2021d).

---

[1] See Blaha (2022c) for example as well as our earlier 2022 books.

2. Primordial Substance (Element)

The possibility of a primordial substance or element has been a subject of interest for about 2500 years. The Pre-Socratic Greeks were the primary investigators in Western Civilization.[2] They assumed that one or more elements were the essence of all things. Choices ranged from Fire to Water to Air and Earth or combinations thereof.

In the late 19th Century Boltzmann championed an atomic origin of all matter based on Statistical Mechanics although the concept of atoms appeared as early as Greek times.

In the 20th Century the concept of a primordial substance was pushed forward to elementary particles and then to the fundamental elementary particles: quarks and leptons together with vector bosons, Higgs particles and gravitons.

An attempt to view the fundamental elementary particles as primordial substances must confront the fact of there are so many and also the fact of the transformations amongst them. A more fundamental problem is the role of Quantum theory in removing their substantiality. We describe particles with Quantum wave functions, which are just as insubstantial as the ether of the 19th Century.

It is perhaps more reasonable to view the wave particles as dynamic constructs causing the apparent evolution of matter. Concomitant with that view is the view of particles as manifestations of space—governed by the essential feature of space: a set of dimensions. With this view we may assert the fundamental particles are embodiments of dimensions encapsulated in dynamics with their lifetimes determined by dynamical conservation laws (and partial conservation laws).

Thus Dimension becomes the primordial substance. Quantum dynamics embodies the particles making them exist. From where does the solidity of matter arise? Solely from the interaction forces of dynamics: primarily from Electromagnetism in its various forms. And also Gravity.

Quantum theory makes matter insubstantial. Dimension defines the forms that matter may take. Dimension specifies the interactions of matter. Dimension leads to our view of matter as substantial. George Berkeley and others proposed a theory in the 1700s called *Immaterialism* (Subjective Idealism) that denied the existence of material substances and suggested that all objects were ideas perceived by the human mind. Dr. Johnson replied with a famous riposte: "I refute it thus" as he kicked a rock.

HyperCosmos theory leads to Dimension being the primordial substance from which all things are constructed. Matter objects are dimensions encapsulated in mathematical dynamical garb.

Physically, The One Dimension is the closest thing to a pure Idea. Matter's solidity is the result of interactions. *Interactions are embodied mathematical ideas.*

---

[2] See G. S. Kirk, J. E. Raven, and M. Schofield, *The Presocratic Philosophers* Cambridge University Press, 1984).

The one primordial dimension is the symbol of an idea that is the portal to the transcendent. What lies beyond is a subject for future study. In chapter 4 we raise the possibility that the One Dimension may be decomposed into a "dust" that becomes the ProtoCosmos space when arranged into a straight line. The Cosmos is thus reduced to structured Dust that may be viewed as Idea.

2.1 Fundamental Particles

In HyperCosmos theory the fundamental fermions of a space's universe are in a one to one relation with the dimensions of its dimension array. Each dimension yields a fermion! Particles and dimensions are inextricably linked.

Every fundamental particle is directly or indirectly composed of dimension(s). The mathematics of its dynamics gives it "solidity" and the possibility of assembly into concrete (macroscopic) objects.

The HyperCosmos dimension array structure maps directly to the spectrum of particles of each space and to the set of particle interactions. The interactions in each space determine the dynamical evolution of its universes.

2.2 Historical Trend of Primordial Elements

The historical trend of Physical primordial elements is

Fire-Earth-Air-Water → Atoms → Quarks, Leptons, … → Dimension

There is a clear conceptual trend present: from the material to the symbolic (Dimension). Physics is proceeding to shed the trappings of matter in favor of the ethereal in the form of structured dimensions with interactions.

Creation is a realization of the symbolic (idea) in the Cosmos and our material universe.

3. Origin of HyperCosmos Dimensions

3.1 Origin of Dimensions in the HyperCosmos

There are several approaches to the definition of the dimensions of the spaces of the HyperCosmos. One approach treats dimension as a primitive quantity. Another approach derives dimensions from features of fermion wave functions. A third approach creates dimensions from the dynamics of the UniDimension ProtoCosmos.

3.2 Dimensions by Fiat

One may take the dimensions of a space to be primitive quantities in the manner of the above passage. Then one can assume the HyperCosmos spectrum of ten spaces (Fig. 3.1) and require the space-time dimensions and the dimensions of the irreducible representations of the internal symmetry groups of each space to be specified by the space's dimension array. The dimension array for the Blaha number 7 space (of our universe) then maps directly to the fermion spectrum and the set of symmetry group irreducible representations of the Unified SuperStandard Theory (UST).

3.3 Derivation of HyperCosmos Dimensions from the Form of Fermion Wave Functions

Free fermion wave functions in the Two-Tier formulation may be written as

$$\psi_1(x) = \Sigma_{\alpha,s}[b_{1\alpha s}u_{\alpha s}f_\alpha(x) + d^\dagger_{1\alpha s}v_{\alpha s}f_\alpha{}^*(x)] \quad (3.1)$$
$$\psi_2(x) = \Sigma_{\alpha,s}[b_{2\alpha s}u_{\alpha s}f_\alpha(x) + d^\dagger_{2\alpha s}v_{\alpha s}f_\alpha{}^*(x)] \quad (3.2)$$

plus Hermitean conjugates with momenta designated by α. *The creation and annihilation operators, the b's and d's, may be viewed as representing dimensions.* This point is made particularly clear[3] by noting General Relativistic transformations transform the 16 creation/annihilation operators in our universe:

$$(b_{1\uparrow}, b_{2\uparrow}, b^\dagger_{1\uparrow}, b^\dagger_{2\uparrow}, \; b_{1\downarrow}, b_{2\downarrow}, b^\dagger_{1\downarrow}, b^\dagger_{2\downarrow}, \; d_{1\uparrow}, d_{2\uparrow}, d^\dagger_{1\uparrow}, d^\dagger_{2\uparrow}, \; d_{1\downarrow}, d_{2\downarrow}, d^\dagger_{1\downarrow}, d^\dagger_{2\downarrow}) \quad (3.3)$$

to linear combinations such as

$$b_{1\beta s} = \Sigma_{\alpha,x,s}\,(g_\beta, f_\alpha)\; u^\dagger_{\beta s}u_{\alpha s}\,(c_{11s}b_{1\alpha s} + c_{12s}b_{2\alpha s} + C_{11s}b^\dagger_{1\alpha s} + C_{12}b^\dagger_{2\alpha s} + $$
$$+ c'_{11s}d_{1\alpha s} + c'_{12s}d_{2\alpha s} + C'_{11s}d^\dagger_{1\alpha} + C'_{12s}d^\dagger_{2\alpha}) \quad (3.4)$$

with similar forms for the other transformed operators (as well as transformed vacua.) General Relativistic transformations T of the form of eq. 3.4 have a 16×16 form.

---

[3] S. Blaha, Phys. Rev. **D17**, 994 (1978); S. Blaha, "The Local Definition of Asymptotic Particle States", IL Nuovo Cimento **49A**, 35 (1979).

3.3.1 Our N = 7 Space Universe

We now introduce internal symmetries in combination with General Relativity in transformations T by promoting creation/annihilation operators to have sedenion[4] features as in Blaha (2022a). We symbolize the sedenion part with an index number σ. Thus eqs. 3.1-3.2 become

$$\psi_{1\sigma}(x) = \Sigma_{\alpha,s}[b_{1\alpha s\,\sigma}u_{\alpha s}f_\alpha(x) + d^\dagger_{1\alpha s\sigma}v_{\alpha s}f_\alpha{}^*(x)] \tag{3.5}$$

$$\psi_2(x) = \Sigma_{\alpha,s}[b_{2\alpha s\sigma}u_{\alpha s}f_\alpha(x) + d^\dagger_{2\alpha s\sigma}v_{\alpha s}f_\alpha{}^*(x)] \tag{3.6}$$

The combined General Relativistic – Internal Symmetry transformation T now has a 256 × 256 form. We can view a special case of the transformation as transforming a 256-vector with the form

$$y = (S_0, S_1, \ldots, S_{15}) \tag{3.7}$$

where

$$S_i = (0, 0, \ldots\, y_i, 0, 0, \ldots) \tag{3.8}$$

for i = 0, 1, 2, …, 15 where y_i represents the ith entry in eq. 3.3 above.

We have called the reference frame of a representation of the form of eqs. 3.7-3.8 a *Fundamental Reference Frame (FRF)*. By having only 16 non-zero items it is analogous to the rest frame of a particle. A transformation T maps 16 items to 256 items. We abstract this transformation into a transformation of dimensions from 16 dimensions to 256 dimensions. In Blaha (2021a) we used this 256 vector, when transformed to a square array, as the equivalent of the sets of fermions and symmetry group representation dimensions in the UST.

Thus we generate a dimension array from the form of creation and annihilation operators suitably generalized to have a sedenion symmetry index for the lowest spin free fermion operators of a space. In this case the space was that of our 4 dimension universe. We introduce General Relativistic – Symmetry transformations that may map dimensions from a FRF to a full 256 dimension array for HyperCosmos N = 7 space.

3.3.2 Dimensions For the Universe of Space N

In the general case of a space of Blaha Number N we may also define dimensions, transformations and an FRF. To determine the number of dimensions in the space N dimension array we begin by noting the free fermion wave function of that space has $2^{r/2\,-\,1}$ spin values with the result that there are $2^{r/2\,+2}$ b's and d's. We generalize the b's and d's to have a corresponding index σ with $2^{2n\,-2}$ values for each creation/annihilation operator where n is the Cayley number corresponding to N.[5] The number of b's and d's increases by a factor of

$$2^{r/2+2} = 2^{2n\,-2} \tag{3.9}$$

---

[4] Sedenions have 16 dimensions.

[5] This procedure is discussed in detail in section 4.1 later.

$(b_{1\uparrow\sigma}, b_{2\uparrow\sigma}, b^{\dagger}_{1\uparrow\sigma}, b^{\dagger}_{2\uparrow\sigma}, b_{1\downarrow\sigma}, b_{2\downarrow\sigma}, b^{\dagger}_{1\downarrow\sigma}, b^{\dagger}_{2\downarrow\sigma}, d_{1\uparrow\sigma}, d_{2\uparrow\sigma}, d^{\dagger}_{1\uparrow\sigma}, d^{\dagger}_{2\uparrow\sigma}, d_{1\downarrow\sigma}, d_{2\downarrow\sigma}, d^{\dagger}_{1\downarrow\sigma}, d^{\dagger}_{2\downarrow\sigma})$ (3.9a)

Thus the total number of enhanced b's and d's is now

$$d_{dN} = 2^{r/2+2} \cdot 2^{2n-2} = 2^{r+4} = 2^{4n-4} \tag{3.10}$$

as shown in Fig. 3.1. Eq. 3.10 gives the total number of dimensions in the space's dimension array.

Thus dimensions are generated in the HyperCosmos from a generalization of the free fermion creation/annihilation operators. We are familiar with space-time dimensions. The creation/annihilation operators define additional dimensions "in momentum space" that unite to form the dimension arrays of HyperCosmos spaces. Note that the dimension arrays describe the "interior" of spaces. Within those spaces dimensions are allocated for the internal space-time of universes defined by the spaces. See Blaha (2022f) and (2022g) for details.

3.4 Dimensions from the Dynamics of the UniDimension ProtoCosmos

The UniDimension ProtoCosmos model[6] is based on a one dimension straight line x composed of primitive dimensions extending from 0 to ∞. The model is based on a generalization of a one dimension Dirac equation with a scaling 1/x potential that leads to a geometric energy spectrum. Defining

$$\psi(x, u) = \psi_1(x)\psi_2(u) \tag{3.11}$$

in

$$[p + m + V(x) + M + i\gamma^{\mu}\cdot\partial/\partial u^{\mu}|_{r = f(E)}]\psi(x, u) = H\psi(x, u) = E\psi(x, u) \tag{3.12}$$

we find the equation separates into two equations

$$[p + m + V(x)]\psi_1(x) = E\psi_1(x) \tag{3.13}$$

$$[\, i\gamma^{\mu}\cdot\partial/\partial u^{\mu}|_{r = f(E)} + M]\psi_2(u) = 0 \tag{3.14}$$

where additional coordinates u of a universe with a Dirac-like interaction term $i\gamma^{\mu}\cdot\partial/\partial u^{\mu}|_{f(E)}$ appear for each energy level E and where

$$f(E) = r = \ln_2[(m - E_N)/16] \tag{3.15}$$

The energy levels determined by eq. 1/13 are:[7]

$$E_N \cong m - d_{dN} \tag{3.16}$$

where

$$d_{dN} = 2^{22 - 2N}$$

[6] See Blaha (2922g) and (2022f) for details.
[7] Using a form agreeing with the HyperCosmos sets of dimension arrays d_{dN}.

where N is an integer.

The solutions of eq. 3.14 (in the PseudoQuantum form) are:[8]

$$\psi_{iMN\alpha}(\mathbf{x}, t, \mathbf{u}, t_u) = \psi_{1iN}(x)\,\psi_{2iMN\alpha}(u)$$

where

$$\psi_{iMN\alpha}(\mathbf{x}, t, \mathbf{u}, t_u) = \psi_{1iN}(x)\,\Sigma_s \int d^{r-1}q\, \mathfrak{N}(q)[u(q, s)b_{iMN\alpha}(q,s)e^{-iq\cdot u} + v(q, s)d_{iMN\alpha}(q,s)^{\dagger}e^{iq\cdot u}] \quad (3.17)$$

$$\psi_{2iMN\alpha}(\mathbf{u}, t_u) = \Sigma_s \int d^{r-1}q\, \mathfrak{N}(q)[u(q, s)b_{iMN\alpha}(q,s)e^{-iq\cdot u} + v(q, s)d_{iMN\alpha}(q,s)^{\dagger}e^{iq\cdot u}] \quad (3.18)$$

for i = 1, 2 where $\mathfrak{N}$ is a normalization constant, where m is the mass in the ProtoCosmos space and M is the mass-energy in the universe space of Blaha number N, r is the number of universe space-time dimensions, and α ranges from 1 through d_{dN}. We assume $q^2 = M^2$ in the universe space. Note $u = (u^0, \mathbf{u})$, and $t_u = u^0$.

The solution eq. 3.22 gives a space fermion of the type described in Blaha (2022c). The fermion is the lowest spin fermion in the space. The range of the symmetry index α from 1 through d_{dN} gives a dimension array of d_{dN} elements that can be subdivided into a d_{cd} by d_{cd} square array of the type seen in the HyperCosmos spaces spectrum. Universe states can be created from sums over α of the wave functions $\psi_{iMN\alpha}$ $(\mathbf{x}, t, \mathbf{u}, t_u)$.

Thus the UniDimension ProtoCosmos model generates dimensions in dimension arrays for each HyperCosmos space. Dimensions originate in the one dimension line in the ProtoCosmos space and subsequently in the energy spectrum fermion wave functions.

In Blaha (2022g) we suggested that the One Dimension origin of everything can be decomposed to generate the dimension straight line of the ProtoCosmos bringing the theory full circle with One Primordial Dimension. The primordial substance is Dimension.

---

[8] We separate the differential equation to obtain the solution in eq. 2.22. In chapter 6 of Blaha (2022c) we form a combined solution for the universe part and the internal universe parts. Physically, the solution eq. 2.22 reflects the separation of the one dimension ProtoCosmos from the subsidiary universes.

3-A. Summary of the HyperCosmos and Unified SuperStandard Theory (UST)

3-A.1 Synopsis of the HyperCosmos and the Unified SuperStandard Theory

Three years ago this author found that the Unified SuperStandard Theory, which was developed previously over a period of years preceding 2019, had a fundamental fermion spectrum and a set of symmetries very similar to a universe theory based on hypercomplex coordinates. Subsequently this author developed, and refined, a hypercomplex space theory for our universe, and a general theory of ten hypercomplex spaces that could support other universes of different space-time dimensions such as the Megaverse (Multiverse). This theory became the HyperCosmos theory.

The HyperCosmos theory has evolved to provide a description of the structure of our universe and other possible universes. We were able to base the HyperCosmos theory on a more fundamental theory called the UniDimension ProtoCosmos theory,[9] which had a wider spectrum of spaces. It included the author's previously considered ten spaces of the HyperCosmos[10] and added infinity of other spaces of fractional dimension generated as an accumulation point of the UniDimension ProtoCosmos spectrum of spaces. We called this additional set of spaces the Limos[11] (after the Lilliputians of *Gulliver's Travels*.)

Equally importantly, we found that one could base the spectrum of fundamental particles and symmetries of each space on a much smaller spectrum of particles and symmetries in a Fundamental Reference Frame (FRF).[12] The spectrum of particles and symmetries in each FRF replicate to form the full spectrum in each HyperCosmos space. The replication is implemented by a General Relativistic transformation[13] in a higher space partner of each space. See Fig. 3-A.1 below. For example, the space of our universe has a 12 dimension higher unification space. Remarkably the unification of the symmetries of our space is implemented in the 12 dimension higher space. This higher space has the overall features of a SuperString F-Theory of 12 dimensions with two time dimensions.

As a result we have unification for any universe of any of the ten HyperCosmos spaces. Each universe of each space type is endowed with a higher dimension unification space that generates its dimension array as a Grand Unified Theory (GUT), and a set of General Relativistic transformations providing Unification. Each universe

---

[9] See Blaha (2022e).

[10] A space defines a space-time and internal symmetry dimensions. A universe is an instance of a space that is created with an initial energy. Each universe has a dimension array containing the fundamental representation dimensions of its internal symmetries and space-time.

[11] Blaha (2022d).

[12] Blaha (2022c).

[13] The General Relativistic transformations are implemented within the author's GiFT formulation of Quantum Field Theory. GiFT combines the author's PseudoQuantum and Two-Tier formulations. See Blaha (2021j).

also has an FRF that reduces its dimension array to a much smaller number of dimensions and thus a much smaller number of fundamental fermions and a much smaller number of symmetries. An analysis of these smaller sets of symmetries and dimensions reduces them to constituent units whose structure is determined by two time coordinates' sublight and superluminal splittings in the space's associated higher space.[14] See chapter 5 and Fig. 3-A.2 below. Our universes' space has a 12 dimension space-time with two time coordinates—equivalent to F-Theory space.

THE HYPERCOSMOS SPACES SPECTRUM

Blaha Space Number $N = O_s$	Cayley-Dickson Number n	Cayley Number d_c	Dimension Array column length d_{cd}	Dimension Array Size d_{dN}	Space-time-Dimension r	Higher Space-time r' Source of d_{dN} r'	Higher Space Array d_{dN}'
0	10	1024	2048	2048^2	18	40	2^{44}
1	9	512	1024	1024^2	16	36	2^{40}
2	8	256	512	512^2	14	32	2^{36}
3	7	128	256	256^2	12	28	2^{32}
4	6	64	128	128^2	10	24	2^{28}
5	5	32	64	64^2	8	20	2^{24}
6	4	16	32	32^2	6	16	2^{20}
7	**3**	**8**	**16**	$\mathbf{16^2}$	**4**	**12**	$\mathbf{2^{16}}$
8	2	4	8	8^2	2	8	2^{12}
9	1	2	4	4^2	0	4	2^{8}

Figure 3-A.1. The HyperCosmos spaces spectrum related to a higher space space-time dimension and its dimension array.

3-A.1.1 Unified SuperStandard Theory

The hypercomplex theory that ultimately emerged began as an attempt by this author in autumn, 2019 to find a deeper basis for his Unified SuperStandard Theory (UST). It became clear that a hypercomplex theory could be the basis of UST. Since then the author developed Octonion Cosmology based on a study of fermion wave function spin in higher space-time dimensions.[15] In the autumn, 2021 the author found that a consideration of General Relativistic transformations of fermion creation and annihilation operators in PseudoQuantum Field Theory led to a much simpler, much improved spectrum of spaces. See Fig. 3-A.1. This effort was the origin of HyperCosmos Theory[16] with a clear relation to fermion creation/annihilation operators. Since then the author has considered the possibility of fractional particles with a lattice-like structuring of fundamental particle interiors, and also connections to partons and Dual Resonance Models (Veneziano amplitudes).

---

[14] Blaha (2022c) and earlier books.
[15] Blaha (2020x).
[16] Blaha (2021i) through (2022c).

The UST has a dimension array that maps to a fermion spectrum and to a symmetry groups spectrum.[17] The maps are well-defined by the hypercomplex structuring of the set of HyperCosmos spaces. Figs. 3-A.3, 3-A.4 and 3-A.5 are for our universe and appear in QUeST which corresponds to the Blaha number N = 7 HyperCosmos space. They separate fermions and symmetries in such a way as to raise the question of the role of symmetry breaking in the Standard Model. The structuring is not dependent on coupling constants and gauge vector boson masses. See chapters 4 and 5 for details related to Figs. 3-A.3 – 3-A.5.

<table>
<tr><td>t_{01}:</td><td colspan="2">sublight
NORMAL
U(4)
8</td><td colspan="2">superluminal
DARK
U(4)
8</td></tr>
<tr><td>t_{02}:</td><td>sublight</td><td>superluminal</td><td>sublight</td><td>superluminal</td></tr>
<tr><td></td><td>SU(3)</td><td>SU(2)</td><td>SU(3)</td><td>SU(2)</td></tr>
<tr><td>**Real Dimensions**</td><td>6</td><td>2</td><td>6</td><td>2</td></tr>
</table>

Figure 3-A.2. Separation of symmetries due to two time coordinates. The separation by two light speeds may account for the Standard Model form of the interaction groups. See chapter 9 for details.

[17] The Dark fermions and symmetries are placed on an equal footing in the UST. The Dark sector could be eliminated if the HyperCosmos set of spaces was redefined.

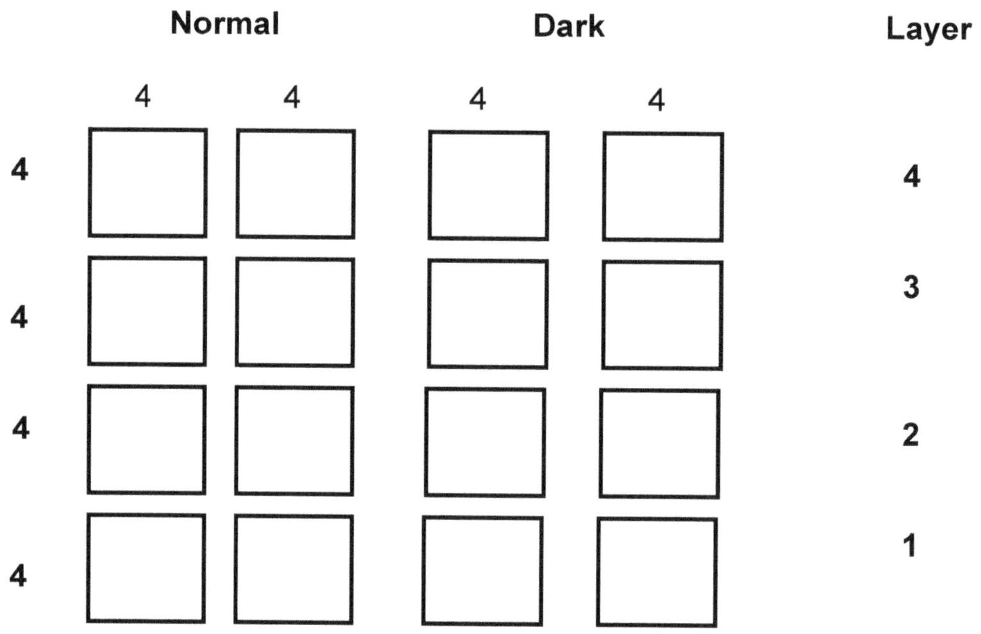

Figure 3-A.3. Block form of a 16 × 16 QUeST/UST dimension array with each block row corresponding to one layer. For the fermion case each block contains four generations. The result is 4 × 4 blocks and a total of 256 fermions.

The Fermion Periodic Table (N = 7)

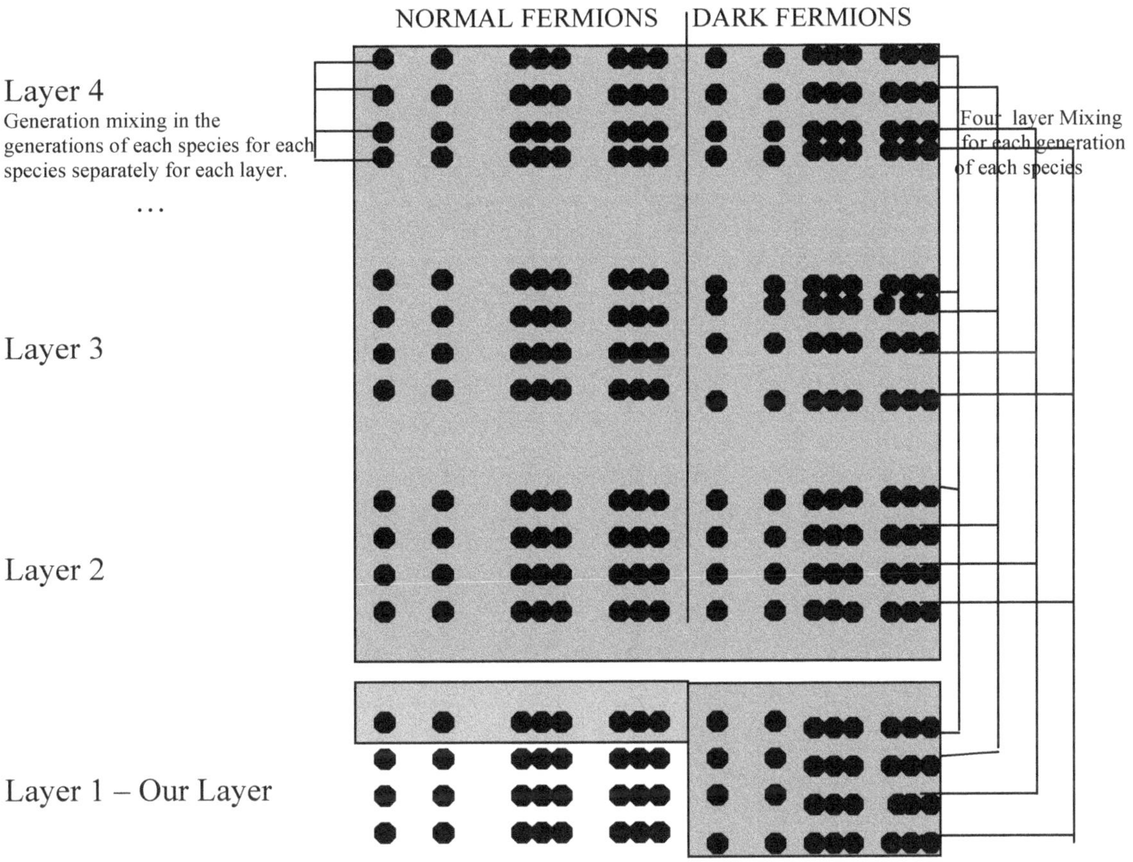

Figure 3-A.4. Fermion particle spectrum and partial examples of the pattern of mass mixing of the Generation groups and of the Layer groups. Unshaded parts are the known fermions, A shaded generation is an additional, as yet not found, 4th generation. The lines on the left side (only shown for one layer) display the Generation mixing within each layer. The Generation mixing occurs within each layer using a separate Generation group for each layer. The lines on the right side show Layer group mixing (for Dark matter) with the mixing among all four layers for each of the four generations individually. There are four Layer groups for Normal matter and four Layer groups for Dark matter.. There are 256 fundamental fermions. QUeST and UST have the same fermion spectrum. Each row is a 16 fermion subblock.

QUeST/UST Symmetries Table

Number of Columns =	4 NORMAL	4	4 DARK	4
Layer 1 4	SU(4)⊗ Generation U(4)	SU(2)⊗U(1)⊗SL(2,C)[18] Layer U(4)	SU(4)⊗ Generation U(4)	SU(2)⊗U(1)⊗SL(2,C) Layer U(4)
Layer 2 4	SU(4)⊗ Generation U(4)	SU(2)⊗U(1)⊗SL(2,C) Layer U(4)	SU(4)⊗ Generation U(4)	SU(2)⊗U(1)⊗SL(2,C) Layer U(4)
Layer 3 4	SU(4)⊗ Generation U(4)	SU(2)⊗U(1)⊗SL(2,C) Layer U(4)	SU(4)⊗ Generation U(4)	SU(2)⊗U(1)⊗SL(2,C) Layer U(4)
Layer 4 4	SU(4)⊗ Generation U(4)	SU(2)⊗U(1)⊗SL(2,C) Layer U(4)	SU(4)⊗ Generation U(4)	SU(2)⊗U(1)⊗SL(2,C) Layer U(4)

Figure 3-A.5. The pattern of symmetries of UST showing the splitting into sections of 4's, 8's, 16's and 32's. Note each layer has its own set of internal symmetries. Normal and Dark sector symmetries are also different. The shaded parts are yet to be found experimentally. As shown, the pattern of symmetries closely matches the QUeST/UST dimension array splittings.The 4 ×4 blocks match the symmetry splits. SU(4) is split into SU(3)⊗U(1) experimentally.The eight SL(2, **C**) groups are transformed into one 4-dimension space-time and seven U(2) Connection Groups.

---

[18] The Lorentz group. $SO^+(1,3)$ in particular

3-A.1.2 The Standard Model Theory

The structure of the Standard Model of Elementary Particles of our universe is present in the Unified SuperStandard Theory. The known fermions appear as unshaded in Fig. 3-A.4. The known symmetries appear as unshaded in Fig. 3-A.5. The unknown fermions seem to be of large mass beyond the reach of current accelerators. The gauge vector bosons of the unknown symmetries also appear to be of large mass.

The Dark sectors were placed in the UST as symmetrical to the normal sectors. Their discovery remains to be done. If the Dark sectors are absent then a revision of the HyperCosmos set of spaces would be necessary. Such a revision would be straightforward and preserve the general characteristics of the HyperCosmos.

The universes of the other higher HyperCosmos spaces have their own USTs, which contain multiple replicates of our UST. For example the next higher space (N = 6) has a Multiverse (Megaverse) instance. It has four replicates of our UST.

For each space above our space, the universe instance has quadruple replicates of the fermions and symmetries of the space/universe immediately below it.

3-A.1.3 Limos and Fractionation

The Limos spaces within the HyperCosmos set of spaces support the fractionation of particles and symmetries as described in Blaha (2022b). Fractionation is analogous to the decomposition of space-time in lattice gauge theory. Fractionation is a more comprehensive decomposition. Particles and symmetries, as well as space-time, are fractionated. In the limit of infinite decomposition quantum theoretic aspects become classical and symmetries become abelian. In this limit a model of the interior "space" of particles using gambols may be constructed.

Gambols appear to be a theoretical basis of partons in Deep Inelastic lepton-nucleon scattering. Also a gambol view of harmonic oscillator models of Dual Resonance phenomena is supported leading to the Veneziano Amplitude. A possibility exists for experimental support through the analysis of ultra high energy quark and lepton collisions for quantum entanglement of gambols.

3-A.2 The Logic of the Derivation of the Unified SuperStandard Theory

The logic of the derivation has several stages. We will specify the stages here and then provide the details in the chapters that follow.

3-A.2.1 The ProtoCosmos Stage

This stage defines a UniDimension ProtoCosmos[19] of one dimension. It has a one dimension Dirac-like equation that generates a bound state spectrum using a one dimension electromagnetic-like potential. The Dirac-like wave function has a factor that is a function of the one dimension coordinate and a factor that is a function of the set of coordinates of an embedded universe. We call the wave function a space wave function since one factor embodies a space. (See Chapter 3.)

The bound state spectrum had the geometric form:

[19] Blaha (2022e).

$$E_N \cong m - d_{dN} \quad (3\text{-A.1})$$

where m is the mass and where

$$d_{dN} = 2^{22-2N} = 2^{r+4} \quad (3\text{-A.2})$$

d_{dN} is the number of components in the dimension array of Blaha number N. Thus the spectrum of HyperCosmos spaces of Fig. 3-A.1 emerges. Note the bound state accumulation point gives the Limos spaces.

We use the relation between Dirac matrix column length γ_C, spin s, and space-time dimension r:

$$\gamma_C = 2^{r/2} = 4s + 2 \quad (3\text{-A.3})$$

implying

$$r = \ln_2(4s + 2)^2$$

where r is the number of space-time dimensions. Then we find the energy E_N satisfies

$$E_N = m - d_{dN} = m - 2\text{^}(\ln_2(4s + 2)^2 + 4) \quad (3\text{-A.4})$$
$$= m - 16(4s + 2)^2$$

or

$$E' = \sqrt{(m - E_N)} = 8 + 16s \quad (3\text{-A.5})$$

is linearly related to the spin of the universe factor of the wave function. This relation is reminiscent of Regge pole trajectories. It leads to the speculation that the higher spaces of the HyperCosmos may be viewed as "resonances" (excitations) of lower spaces. (See chapter 4 of Blaha (2022f). The ProtoCosmos generates the HyperCosmos spectrum of spaces in the UniDimension Model and the Tower Model.

3-A.2.1.1 Generation of Spaces from an Energy

The energy relation implied by eqs. 3-A.1 and 3-A.2 for m = 0 and E" = -E is

$$r = \ln_2 E'' - 4 = \ln_2 E''/16 \quad (3\text{-A.6})$$

If we require r to be an integer then E" must have the form

$$E'' = 2^k = 2^{22-2N} \quad (3\text{-A.7})$$

for k = 4, 5, 6, … or N = 0, 1, 2, … 9. Thus we obtain part of the HyperCosmos spectrum of spaces. If we allow k < 4 values that give r values that are fractional powers of 2 then we obtain the Limos part of the HyperCosmos spectrum.

Thus one may envision a zero space-time dimension ProtoCosmos where a point (the Tower ProtoCosmos) has a series of allowed values for E" and a set of spaces is generated of the form of the HyperCosmos spectrum. Chapter 4 shows each space has an energy E" and spin s satisfying

$$E'' = 16(4s + 2)^2 \qquad (3\text{-}A.8)$$

as seen above. We define a ProtoCosmos amplitude α(s) with

$$\alpha(s) = \sqrt{E''} = 4(4s + 2) \qquad (3\text{-}A.8a)$$

α(s) has a linear dependence on spin s similar to a Regge pole trajectory. At certain spins we find corresponding HyperCosmos spaces. We specify them as functions of Blaha number N:

$$\alpha(s_N) = \sqrt{E''_N} = 4(4s_N + 2) \qquad (3\text{-}A.8b)$$

in analogy with Regge pole trajectory resonances. See chapter 2 for details.

Since the dimension array column length satisfies

$$d_{cdN} = \sqrt{d_{dN}} = \sqrt{E''_N} = \alpha(s_N) \qquad (3\text{-}A.9)$$

we find

$$d_{cdN} = 16s_N + 8 \qquad (3\text{-}A.10)$$

where s_N is the Blaha space N wave function spin.

The ProtoCosmos then has a spaces trajectory similar to Regge pole trajectories. If we allow N to be continuous then we obtain a continuous spectrum of space-time dimensions r and non-integer-based HyperCosmos spaces interspersed within the integer-based spaces in Fig. 3-A.1.

These considerations raise the possibility of a deeper synthesis of the ProtoCosmos.

3-A.2.2 The HyperCosmos Stage

The spectrum of spaces that we found for the HyperCosmos invites the view that it corresponds to an inverted spectrum of states where d_{cd} (of Fig. 3-A.2) plays a role analogous to (negative) energy, and the Blaha space number $N = 0_s$ plays the role of an energy level index. First we look at the inverted HyperCosmos spaces spectrum in Fig 3-A.6 below. It reveals a Hydrogen-like energy spectrum. Note the energy accumulation point at zero as $N = O_s \rightarrow \infty$. Note also the geometric scaling behavior of the energy d_{cd}. Fig. 3-A.6 shows the inverted spectrum plotted as an energy spectrum.

Inverting the energy scale (and making the energy sign positive) gives the HyperCosmos space spectrum. Chapters 2 and 3 of Blaha (2022f) developed the ProtoCosmos. Then chapter 4 shows it has an amplitude – spin relation that gives the space spectrum a Regge trajectory analogue. Chapter 5 of Blaha (2022f) describes the wave function solutions of the UniDimension ProtoCosmos that each contains a different universe part.

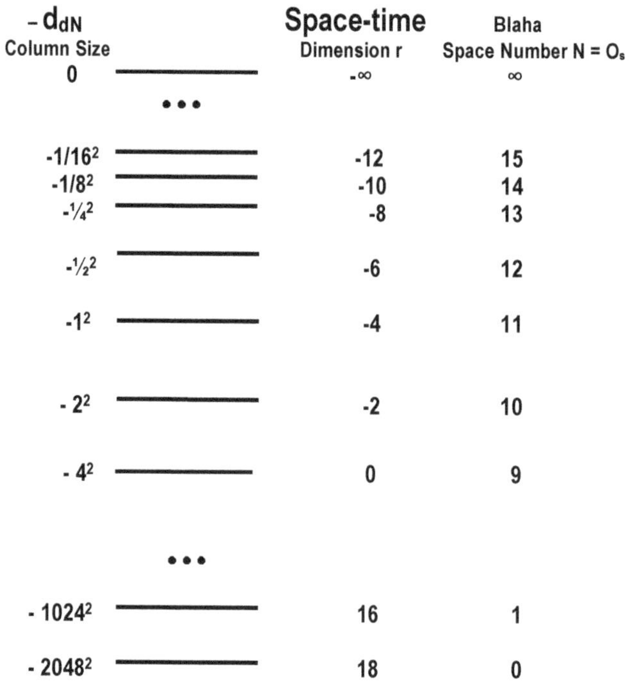

Figure 3-A.6. HyperCosmos spectrum plotted as an Inverted energy spectrum with E = – d_{dN}. See Fig. 3-A.3 below.

3-A.2.3 The QUeST Stage

Our universe is based on the N = 7 HyperCosmos space. QUeST was its original Model.

3-A.2.4 The UST Stage

QUeST originated in the Unified SuperStandard Theory (UST) developed by the author a few years ago. The author found QUeST and UST had almost the same set of features.

3-A.2.5 The Standard Model Stage

The UST originated as a generalization of the Standard Model. It added Two-Tier quantization to avoid divergences in perturbation theory. Two-Tier theory also enables higher dimension quantum field theories to avoid divergences.

The UST also embodied the author's PseudoQuantum Theory develop by the author in the 1970s. PseudoQuantum Theory led to the HyperCosmos and its use of Fundamental Reference Frames for unification of General Relativity and Internal Symmetries.

Appendix 3-B. Experimental Support for the HyperCosmos and UST

There is evidence for the ProtoCosmos and the HyperCosmos. It is not direct evidence since we are bound by our universe and its elapsed lifetime. It is evidence that indirectly supports their existence and their structure.

1. Direct Generation of the HyperCosmos spectrum from a UniDimension ProtoCosmos.

2. Relic Structure in Internal Symmetries and the Fundamental Fermion Spectrum. The Standard Model internal symmetry SU(2)⊗U(1)⊗SU(3) is shown to appear in HyperCosmos/QUeST as a natural separation not requiring symmetry breaking Higgs fields. QUeST introduces additional symmetries that add to those of the Standard Model in a direct way—again without Higgs bosons. Higgs breaking is restricted to ElectroWeak-like sectors.

3. *The separation of fermions and symmetries into superluminal and subluminal parts in the Fundamental Reference Frame leads to the fundamental fermion spectrum, to the set of internal symmetries, and to the understanding of the relation of the Higgs Mechanism and Strong Interaction Confinement (the Higgs-Confinement Dichotomy).*

4. The Dual Resonance Model of hadron scattering is *directly* based on the HyperCosmos using fractionated hadrons. This approach is an alternative to String Theories. This approach supports hadronic Veneziano amplitudes.

5. The Parton Model of Deep Inelastic Lepton-Nucleon Scattering may be viewed as based on fractionated particles. One finds that it "explains" quasi-free partons, the apparent lack of strong interactions between partons thus giving parton freedom, and scaling in x, as well as other features.

6. The HyperCosmos spaces spectrum can be cast into a form similar to the Regge trajectories of scattering amplitudes.

7. The fractionated particle formalism has lattice theory as a restricted special case.

8. The HyperCosmos leads to a deep unification based on its Fundamental Reference Frame formalism. For example, in the case of our universe, there is a unified 12 space-time dimension formalism that generates the unified General Relativity and Internal Symmetries of our universe.

Additional circumstantial evidence appears in the 42 dimensions of the Full HyperUnification space and 88 dimensions of the UltraUnification space as discussed in the Introduction and chapter 9.

4. Fundamental Reference Frames (FRF), Dimensions and Replicates

The combination of General Relativity and Internal Symmetries in the transformations from an FRF fills the dimension array of a HyperCosmos space. The dimensions in the array play several roles: they map to fundamental fermions; alternately they map to sets of internal symmetry fundamental representation dimensions; and also they map to sets of Higgs and other scalar bosons. See Blaha (2022g) for more details. This chapter describes the transformation from a set of dimensions in the FRF to the full set of dimensions in a universe. The transformation also applies to core sets of fermions andof symmetry interactions. These transformations create replicates of the core sets.

The following chapters extend the transformations to generate dimensions from a single dimension and extend the set of dimensions of each space to its HyperUnification space.

4.1 The Fundamental Reference Frame (FRF) as a "Rest Frame" for Reference Frames

A common feature of calculations in Physics is to choose a reference frame which facilitates computation. Common choices are the rest frame and the center of mass frame. In this chapter we find a Fundamental Reference Frame (FRF) (actually an infinite set of such frames) that plays a similar role for coordinate reference frames. We describe how to determine a Fundamental Reference Frame for each of the ten spaces (universes) of the HyperCosmos.

We suggest that a set of transformations exist that map from an FRF to a "normal" static reference frame. We are motivated in this endeavor by the observation made in previous books (and later) that the fermion particles of a space (universe) appear to be expressible as sets of replicated fermions. The replication of fermions is generated by transformations of an FRF set of fermions to a set of replicates of the FRF set.

The discussion begins with the results of section 3.3.2. In the N = 7 space of our universe the 16 creation/annihilation b's and d's operators are

$$B = (b_{1\uparrow}, b_{2\uparrow}, b^{\dagger}_{1\uparrow}, b^{\dagger}_{2\uparrow},\ b_{1\downarrow}, b_{2\downarrow}, b^{\dagger}_{1\downarrow}, b^{\dagger}_{2\downarrow},\ d_{1\uparrow}, d_{2\uparrow}, d^{\dagger}_{1\uparrow}, d^{\dagger}_{2\uparrow},\ d_{1\downarrow}, d_{2\downarrow}, d^{\dagger}_{1\downarrow}, d^{\dagger}_{2\downarrow}) \qquad (4.1)$$

in the PseudoQuantum formulation for the lowest spin fermion. A General Relativistic transformation T transforms B to a linear combination of b's and d's denoted B' as shown in our earlier work:

$$B' = TB \qquad (4.2)$$

The transformation T has the form of a 16 by 16 matrix.

We now wish to introduce an Internal Symmetry part with the goal of a complete unified theory. We add an index σ to each of the b's and d's where σ ranges from 1 through

$$2^{r/2+2} = 2^{2n-2} = 16 \tag{4.3}$$

where the space-time dimension is r = 4 and the Cayley number n = 3 for the N = 7 space of our universe. The b's and d's now number 256.

$$(b_{1\uparrow\sigma}, b_{2\uparrow\sigma}, b^{\dagger}_{1\uparrow\sigma}, b^{\dagger}_{2\uparrow\sigma}, b_{1\downarrow\sigma}, b_{2\downarrow\sigma}, b^{\dagger}_{1\downarrow\sigma}, b^{\dagger}_{2\downarrow\sigma}, d_{1\uparrow\sigma}, d_{2\uparrow\sigma}, d^{\dagger}_{1\uparrow\sigma}, d^{\dagger}_{2\uparrow\sigma}, d_{1\downarrow\sigma}, d_{2\downarrow\sigma}, d^{\dagger}_{1\downarrow\sigma}, d^{\dagger}_{2\downarrow\sigma}) \tag{4.4}$$

The choice of the number of σ components, 16 in this case, is motivated by the goal of symmetry between the number of spin determined b and d components and the number of Internal Symmetry determined components—16 in both cases in this space. .

We now note the number of b's and d's are equal to the number of elements d_{dN} in the N = 7 dimension array of Fig. 1.1.

$$d_{dN} = 2^{r/2+2} \cdot 2^{2n-2} = 2^{r+4} = 2^{4n-4} = 256 \tag{4.5}$$

The enhanced transformation of the b's and d's B_E has the form:

$$B_E' = T_E B_E \tag{4.6}$$

where T_E is a 256×256 array.

T_E is a General Relativistic – Internal Symmetry transformation.

4.1.1 HyperUnification Space

We now consider a related space of our N = 7 space. This space has transformations of the form of eq. 4.6 where the transformations are purely General Relativistic. Our goal for this space, which we call the HyperUnification space for our universe, is to unify space-time and internal symmetries. We wish it to have the form of HyperCosmos spaces.

In Blaha (2022e) and (2022g) we presented a unification for our universe based on a map to a higher space-time dimension r' HyperCosmos space called HyperUnification space.[20] This HyperUnification space has a General Relativity that generates transformations that map to forms of the dimension array $d_{dN'}$ where N' is the Blaha number of the HyperUnification space. The HyperUnification space and the "lower" space have space-time dimensions related by:

---

[20] See Blaha (2022d) and02022g) for more detail. The higher dimension space may be one of the ten HyperCosmos spaces or it may be another HyperCosmos-like space. The unification spaces have space-time dimensions ranging up to 40 dimensions. See Fig. 4.1.

$$d_{dN} = 2^{22-2N} = 2^{r+4} \tag{4.7}$$

$$r' = 2r + 4 \tag{4.8}$$

based on the relation of the r dimension transformation and the r' dimension transformation.

<u>General Justification</u>

To obtain the dimension array for the r coordinate space-time we can use the r' coordinates space-time of the HyperUnification space. The r' space-time transformation has a $2^{r'/2+2} \times 2^{r'/2+2} = 2^{r+4} \times 2^{r+4}$ transformation array that generates a $2^{r'/2+2} = 2^{r+4}$ coordinates vector from its FRF. This vector maps to the r space-time dimension array d_{dN}.[21] The HyperUnification transformations are General Relativistic transformations of the HyperUnification space. *These transformations embody both the General Relativistic space-time and Internal Symmetries of the lower HyperCosmos space.*

The $2^{r'/2+2}$-vector elements of the HyperUnification space become the elements of the d_{dN} dimension array of the "lower" space.

The defining feature of a HyperUnification space with space-time dimension r' of a space with space-time dimension r is that the number of components in the dimension r' GR transformation vector equals the number of components d_{dN} *in the dimension array of the space of space-time dimension r.*

For example, in our four dimension universe the HyperUnification space has 12 space-time coordinates. Its dimension array has 256×256 = 65,536 elements.[22] It has a 256-vector that becomes the d_{dN} = 256 elements dimension array[23] of our universe with 4 space-time dimensions and 252 internal symmetry dimensions. The HyperUnification space-time dimension procedure yields the r space-time dimension array. *The key role of the HyperUnification space is Unification.*

Note:

We have used the space corresponding to r' space-time dimensions to create the r space-time dimension array. The dimension array in each space-time specifies the number of fermion creation and annihilation operators in that space-time.

The number of elements in the dimension array of the HyperUnification space of space-time dimension r' is:

$$d_{dN'} = 2^{r'+4} \tag{4.9}$$

In terms of r' we find its Blaha number N' is

$$N' = \tfrac{1}{2}(18 - r') = 7 - r \tag{4.10}$$

---

[21] The array is divided into a square d_{dN} array.

[22] The value of the elements is not relevant for our discussions. The fact that they are non-zero in general is the key point.

[23] Again we note the value of the elements is not relevant. They play a symbolic role in being mapped to fermions and mapped to symmetry group dimensions, and so on.

or

$$N = 9 - r/2 \tag{4.11}$$

Note: negative values of N' are allowed for spaces where r > 7 in eq. 4.10.

THE HYPERCOSMOS SPACES SPECTRUM

Blaha Space Number $N = O_s$	Cayley-Dickson Number n	Cayley Number d_c	Dimension Array column length d_{cd}	Dimension Array Size d_{dN}	Space-time-Dimension r	Space-time r' Source of d_{dN} r'	HyperUnification Space Array d_{dN}'
0	10	1024	2048	2048^2	18	40	2^{44}
1	9	512	1024	1024^2	16	36	2^{40}
2	8	256	512	512^2	14	32	2^{36}
3	7	128	256	256^2	12	28	2^{32}
4	6	64	128	128^2	10	24	2^{28}
5	5	32	64	64^2	8	20	2^{24}
6	4	16	32	32^2	6	16	2^{20}
7	**3**	**8**	**16**	$\mathbf{16^2}$	**4**	**12**	$\mathbf{2^{16}}$
8	2	4	8	8^2	2	8	2^{12}
9	1	2	4	4^2	0	4	2^{8}

Figure 4.1. The HyperCosmos spaces spectrum related to a HyperUnification space space-time dimension r' and its dimension array. Note the spaces with r' > 18 are outside the HyperCosmos set of 10 spaces. However they are made to have the same form as the 10 HyperCosmos spaces.

4.2 Contents of Each FRF

The contents of a HyperCosmos space FRF number $2^{r/2+2}$ non-zero elements where r is the space-time dimension.[24] The number of non-zero elements equals the number of b's and d's, 16, before internal symmetries are introduced as in eq. 4.1. After internal symmetries are introduced into the b's and d's as in eq. 4.3, the number of b's and d's is 2^{r+4}. Correspondingly, the number of elements in the dimension array is 2^{r+4}.

Earlier we found 16 b's and d's in our universe's space. These b's and d's become the non-zero elements in our universe's FRF. In general for $r \geq 4$ there are $2^{r/2-2}$ replicates of these 16 non-zero elements in the space's FRF. The remainder of the FRF vector is $2^{r+4} - 2^{r/2+2}$ zeroes. In the space of our universe there are 240 zeroes and 16 non-zero items in the FRF vector.

Justification

A basis for the above result is the demonstration in chapter 5 that the 16 elements of our space's FRF can be split into Internal Symmetry groups based on the two time dimensions in our universe's HyperUnification space (and in its F-Theory analogue).

---

[24] The total number of components in the FRF vector is 2^{r+4} counting non-zero and zero-valued components.

Each HyperCosmos space has a lowest spin fermion wave function with $2^{r/2-2}$ replicates of the b's and d's of eq. 4.1. The replication of b's and d's matches the replication of 16 non-zero element sets in the FRF.

The contents of the FRF's are listed in Figs. 4.2 – 4.4 below in terms of Blaha number N. Fig. 4.7 – 4.10 shows the fermion and internal symmetry group content of the r = 4 (N = 7) FRF.

The non-zero elements in an FRF correspond to fundamental fermions, Internal Symmetry groups, or dimensions.

Contents of Fundamental Reference Frames

N	Number of Physical Elements	Total Number of Elements Generated
0	2^{11}	2^{22}
1	2^{10}	2^{20}
2	2^{9}	2^{18}
3	2^{8}	2^{16}
4	2^{7}	2^{14}
5	2^{6}	2^{12}
6	2^{5}	2^{10}
7	2^{4}	2^{8}
8	2^{3}	2^{6}
9	2^{2}	2^{4}

Figure 4.2. The contents of the ten types of Fundamental Reference Frames for the ten spaces. The elements are fermions or dimensions or symmetry fundamental representation dimensions.

Contents of a Fundamental Reference Frame (FRF) for Fermions

N	Number of Physical Elements	Fermion Content	
0	2^{11}		
1	2^{10}		
2	2^{9}		
3	2^{8}	•	
4	2^{7}	•	
5	2^{6}	•	
6	2^{5}	NORMAL e q-up v q-down e′ q-up′ v′ q-down′ DARK e q-up v q-down e″ q-up″ v″ q-down″	Total: 32 fermions
7	2^{4}	Normal: e q-up v q-down Dark: e q-up v q-down	16 fermions
8	2^{3}	e q-up AND v q-down	8 fermions
9	2^{2}	e q-up OR v q-down	4 fermions

Figure 4.3. Possible contents of five of the ten types of Fundamental Reference Frames for fermions. The primes distinguish different fermions.

The symmetry group contents in the Fundamental Reference Frame are listed for N > 5 below. Note the units of symmetry irreducible representations dimensions are four real-valued dimensions. Irreducible symmetry group units also have four dimensions. The number of units repeatedly doubles as we ascend up to N = 0.

Contents of a Fundamental Reference Frame for Symmetry Group Dimensions

N	Number of Physical Elements	Fermion Content	
0	2^{11}		
1	2^{10}		
2	2^{9}		
3	2^{8}	•	
4	2^{7}	•	
5	2^{6}	•	
6	2^{5}	NORMAL U(4)⊗U(4) DARK U(4)⊗U(4)	Total: 32 real dimensions
7	2^{4}	NORMAL U(4) ⊗ DARK U(4)	16 real dimensions
8	2^{3}	U(4)	8 real dimensions
9	2^{2}	U(2)	4 real dimensions

Figure 4.4. Possible contents of five of the ten types of Fundamental Reference Frames for symmetry dimensions. The groups may undergo a further breakdown after being mapped to a static space-time to give terms like those in Fig. 4.4.

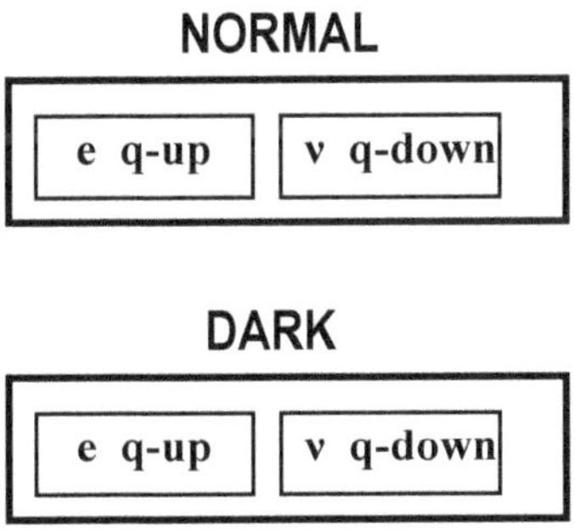

Figure 4.5. The 16 fermions that constitute the set that is replicated 16 times in the 256 fermion spectrum for N = 7 space (universe).

The separation of the 16 fermions into two pairs of four fermions in Fig. 4.5 reflects the general separation of fermions and internal symmetries into 4, 8, 16 32, and 64 parts due to symmetry splitting as we discussed in Blaha (2021b).

The set of symmetry groups in QUeST for our universe also exhibits the same sort of replication seen for fundamental fermions. In the FRF there are 16 dimensions for U(4)⊗U(4) according to Fig. 4.4 for N = 7.

After a transformation, the symmetry groups have 256 real-valued dimensions for a product of the fundamental representations of their component groups. The basic symmetry group set for an N = 7 space is a U(4)⊗U(4) group. After a 16-fold replication the U(4) factors undergo transformations/breakdowns to

U(4) or SU(3)⊗U(1) or SU(2)⊗U(1)⊗SL(2, **C**)[25] or SU(2)⊗U(1)⊗U(2)

as seen below.

4.3 Replicates Generated by Transformations in HyperCosmos Spaces

The replication of the $2^{r/2+2}$ non-zero elements of an FRF is the result of a transformation in the HyperUnification as shown in eq. 4.6. Thus there is a replication of a replication in the FRF. This is visualized in Fig. 4.6.

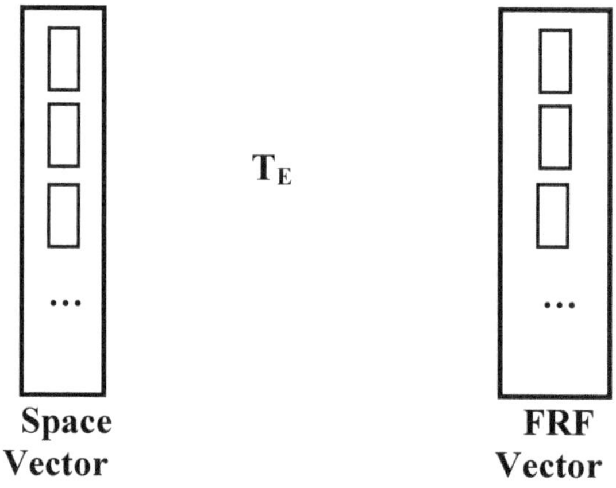

Figure 4.6. Visualisation of the T_E transformation replication pattern in the HyperUnification space and its FRF. The FRF has $2^{r/2-2}$ replicates of the basic 16 dimension set. The space vector has 2^r copies of the basic set 16 dimension set. The transformation generates $2^{r/2+2}$ replicates of the FRF's $2^{r/2-2}$ replicates of the basic 16 dimension set.

The FRF of our N = 7 space has a $2^0 = 1$ copy of the basic set in the FRF vector. The purely General Relativistic HyperUnification transformation generates 16 replicates

[25] We use SL(2, C) to represent $SO^+(1, 3)$.

or 256 non-zero dimensions. These replicates may be visualized in Figs. 4.7 – 4.10 for our universe's QUeST/UST.

Number of Columns =	4 NORMAL 4		4 DARK 4	
Layer 1 4	e 3 up-quarks	ν 3 down-quarks	e 3 up-quarks	ν 3 down-quarks
Layer 2 4	e 3 up-quarks	ν 3 down-quarks	e 3 up-quarks	ν 3 down-quarks
Layer 3 4	e 3 up-quarks	ν 3 down-quarks	e 3 up-quarks	ν 3 down-quarks
Layer 4 4	e 3 up-quarks	ν 3 down-quarks	e 3 up-quarks	ν 3 down-quarks

Figure 4.7. The QUeST 16 × 16 fermion spectrum of our universe tentatively arranged as SU(4)-plets that correspond directly with SU(4) (or SU(3)⊗U(1)) fermions. There are four layers. Each set of 4 fermions has 4 generations matching the number of rows in each layer. This Periodic Table is broken into Normal and Dark sectors.

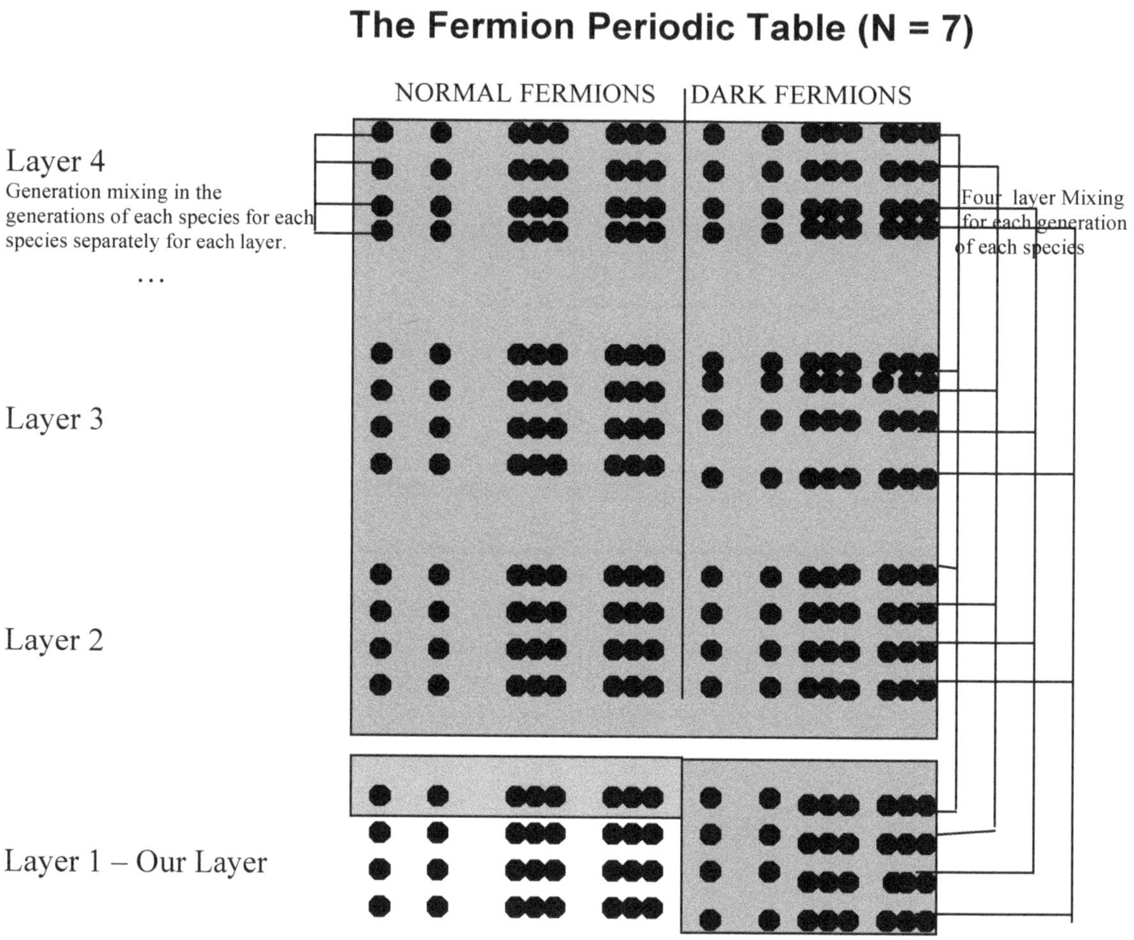

Figure 4.8. QUeST Fermion particle spectrum and partial examples of the pattern of mass mixing of the Generation groups and of the Layer groups. Unshaded parts are the known fermions, A shaded generation is an additional, as yet not found, 4th generation. The lines on the left side (only shown for one layer) display the Generation Group mixing within each layer. The Generation mixing occurs within each layer using a separate Generation group for each layer. The lines on the right side show Layer group mixing (for Dark matter) with the mixing among all four layers for each of the four generations individually. There are four Layer groups for Normal matter and four Layer groups for Dark matter. There are 256 fundamental fermions. QUeST and UST have the same fermion spectrum.

Initially the N = 7 QUeST/UST full set of groups has the form of 16 U(4)⊗U(4) symmetry groups (Fig. 4.3.) Then they are transformed by symmetry breaking to the groups in Fig. 4.7.

	NORMAL	DARK
Layer 1	U(4)⊗U(4) U(4)⊗U(4)	U(4)⊗U(4) U(4)⊗U(4)
Layer 2	U(4)⊗U(4) U(4)⊗U(4)	U(4)⊗U(4) U(4)⊗U(4)
Layer 3	U(4)⊗U(4) U(4)⊗U(4)	U(4)⊗U(4) U(4)⊗U(4)
Layer 4	U(4)⊗U(4) U(4)⊗U(4)	U(4)⊗U(4) U(4)⊗U(4)

Figure 4.9. The "initial" distribution of sets of N = 7 symmetry groups. Each set is distinct and supports interactions only for the corresponding set of fermions (separately for Normal and Dark fermions). *Thus each set of 16 fermion generations has its own quantum numbers and interactions.* Each U(4)⊗U(4) set has a 16 real-valued dimension representation, which are importance when we consider Fundamental Reference Frames.

NORMAL		DARK	
SU(3)⊗U(1) Generation U(4)	SU(2)⊗U(1)⊗SL(2, C) Layer U(4)	SU(3)⊗U(1) Generation U(4)	SU(2)⊗U(1)⊗U(2) Layer U(4)
SU(3)⊗U(1) Generation U(4)	SU(2)⊗U(1)⊗U(2) Layer U(4)	SU(3)⊗U(1) Generation U(4)	SU(2)⊗U(1)⊗U(2) Layer U(4)
SU(3)⊗U(1) Generation U(4)	SU(2)⊗U(1)⊗U(2) Layer U(4)	SU(3)⊗U(1) Generation U(4)	SU(2)⊗U(1)⊗U(2) Layer U(4)
SU(3)⊗U(1) Generation U(4)	SU(2)⊗U(1)⊗U(2) Layer U(4)	SU(3)⊗U(1) Generation U(4)	SU(2)⊗U(1)⊗U(2) Layer U(4)

Figure 4.10. The transformed/broken sets of symmetries in QUeST (UST) and in N = 7 HyperCosmos space.. Note each element has a 16 real dimension representation. This depiction is also evident in QUeST and the UST. The SL(2, **C**) representation has four coordinates.[26]

---

[26] The Lorentz Group $SO^{+}(1, 3)$ is often specified with an SL(2, **C**) representation.

4.3 Portrait of a HyperCosmos Space and its Associated HyperUnification Space

Fig. 4.11 shows the relation between a HyperCosmos space and its related HyperUnification space and their FRFs.

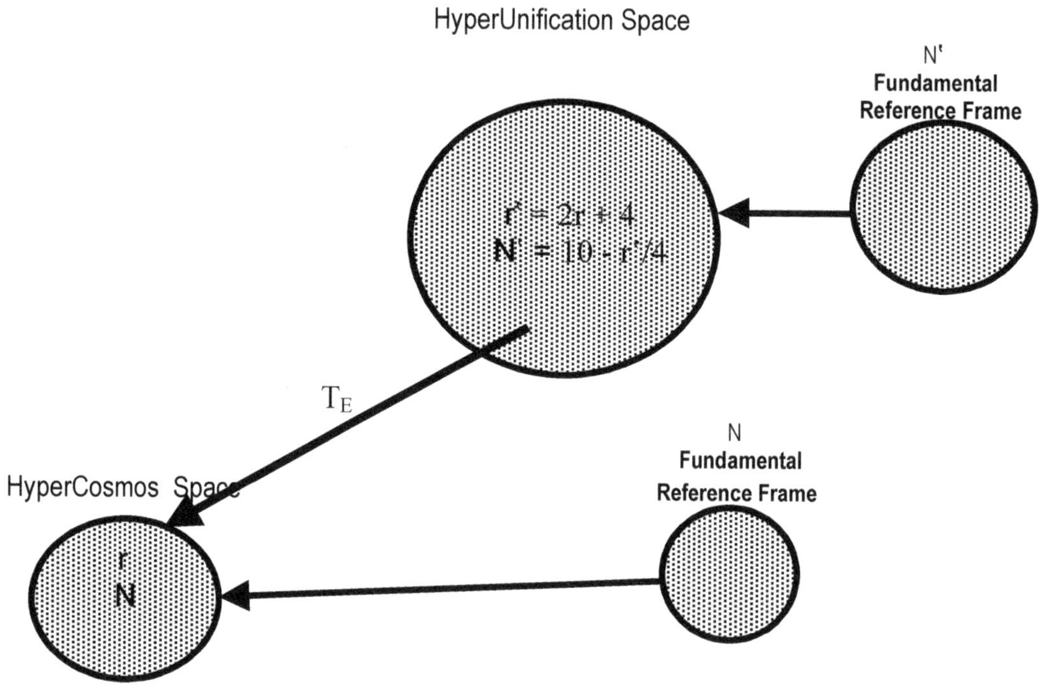

Figure 4.11. Diagram of the relation between a HyperCosmos space and its associated HyperUnification space with Fundamental Reference Frames indicated. The HyperUnification space sets the dimension array d_{dN} for the HyperCosmos space using purely HyperUnification General Relativity. Fundamental Reference Frames are like the rest frames of particles. They reduce the number of elements to a core set of elements. The elements are dimensions or fermions or gauge vector symmetry fundamental representation dimensions.

4.4 Reduction of an FRF's Contents to One Non-Zero Dimension in a HyperUnification Space

A HyperUnification Space for $r \geq 4$ has an FRF with $2^{r/2-2}$ replicates of the 16 non-zero elements in our universe's FRF according to section 4.2. The 16 elements corresponded to the 16 b/\'s and d's of a spin ½ fermion in our universe. The remainder of the FRF vector is $2^{r+4} - 2^{r/2+2}$ zeroes.

Transformations of the HyperUnification space also support the generation of the *entire* set of HyperCosmos dimensions from one dimension in its FRF vector in the HyperUnification Space.

We demonstrate this possibility with a simple example. Assuming the FRF has one non-zero dimension (element) in row 1 of the FRF vector with all other components zero we define a vector with d_{dN} components as

$$V = (a, 0, 0, \ldots) \tag{4.12}$$

and a prototype General Relativistic transformation

$$[T]_{ij} = b_i \ \delta_{i1} + (1 - \delta_{1i})(1 - \delta_{j1})c_{ij} \tag{4.13}$$

where b_i and the c_{ij} are numeric. Transformation T has entries b_i in column 1 and a matrix c_{ij} in the rows and columns for i, j = 2, 3, … . A full set of d_{dN} array dimensions are generated from one FRF dimension (element). The transformation

$$V' = TV \tag{4.14}$$

generates the vector V' with no zero-valued components:

$$V' = (a_1, a_2, a_3, \ldots, a_{d_{dN}}) \tag{4.15}$$

This vector represents the d_{dN} dimension array when it is restructured into a square array. The representation is not numeric since dimensions have no numeric value. Each non-zero value in the dimension array signifies it is an *active* dimension that is used in determining a fundamental fermion spectrum or in determining the dimensions of an irreducible symmetry group representation.

Thus we can generate the complete set of dimensions in each HyperCosmos dimension array from one dimension in its FRF using a HyperUnification space General Relativistic transformation.

4.5 FRFs and Replicates

The $2^{r/2+2}$ non-zero elements in an FRF are generated from one element by a HyperUnification transformation. These elements consist of $2^{r/2-2}$ replicates of the 16 elements of our universe's FRF. Each replicate has different groups from the other replicates.

A further HyperUnification transformation maps the FRF contents to the d_{dN} elements that become a HyperCosmos dimension array when expressed as a square array.[27] This second transformation creates $2^{r/2+2}$ replicates of each of the FRF's $2^{r/2+2}$ elements (the $2^{r/2-2}$ replicates of the 16 elements in the FRF).

Thus the dimension array that emerges consists of 2^{r+4} elements distributed in 2^r replicates of the 16 elements of our universe's FRF. Each of these replicates corresponds to different sets of fundamental fermions or to different sets of irreducible representations of symmetry groups.

---

[27] The pair of transformations may be combined into one transformation from one element to the d_{dN} element vector that becomes the dimension array of the HyperCosmos space.

We see these replicates in the QUeST and UST figures for our universe! Chapter 5 analyzes the 16 elements of our universe's FRF and shows that they reflect separations into sublight and superluminal parts based on the two time dimensions in our 12 space-time dimension HyperUnification space, which is equivalent to F-Theory space. See Figs. 4.7 and 4.10.

5. FRF Contents Determined by Two Time Dimensions in HyperUnification Space

5.1 The Higgs Mechanism – Quark Confinement Dichotomy

A possibly deeper view of the Higgs Mechanism is based on a Dichotomy between the Higgs Mechanism and Quark confinement. The Higgs Mechanism and the confining Strong Interaction are alternatives. The Higgs Mechanism gives masses to particles. Quark Confinement is based on an effective potential energy that prevents quark deconfinement. Both affect particle energies.

The Higgs Mechanism uses a tachyonic scalar particle implementation[28] for symmetry breaking because it works despite it being superluminal. The Strong Interaction has a non-Higgs non-tachyonic confinement mechanism.[29] These mechanisms may be related.

A simple boson example illustrating the dichotomy in the mechanisms starts with the PseudoQuantum scalar field equations:

$$\Box\varphi_1 - m^2\varphi_1 + g\varphi_1{}^4 = 0 \tag{5.1}$$
$$\Box\varphi_2 - m^2\varphi_2 + g\varphi_2{}^4 = 0 \tag{5.2}$$

with tachyonic negative mass squared. Both the type 1 and type 2 fields have the same field equation as we have seen in earlier work.

On the other hand, the prototype Strong Interaction PseudoQuantum scalar field equations:

$$\Box\varphi_1 \pm m^2\varphi_2 = 0 \tag{5.3}$$
$$\Box\varphi_2 = J \tag{5.4}$$

with the quartic interaction dropped for simplicity yield

$$\Box^2\varphi_1 \pm m^2 J = 0 \tag{5.5}$$

Choosing the minus sign to have a positive J source term: results in

$$\Box^2\varphi_1 = m^2 J \tag{5.6}$$

which specifies a confining linear r potential when redone as a gauge field theory.[30]

---

[28] The tachyonic nature is removed by quartic interactions.

[29] See Blaha (2022f) for a detailed discussion of quark confinement and of related MOND-like gravitation.

[30] See Stephen Blaha, "New Framework for Gauge Field Theories", IL Nuovo Cimento **49A**, 113 (1979) for an SU(3) gauge theory version with confinement.

Note the minus sign places eqs. 5.1 and 5.3 on the same footing with the difference that eq. 5.1 is tachyonic but eq.5.3 is not tachyonic due to the appearance of the type 2 field in the eq. 5.3 mass term. "Tachyonicism" is avoided in order to obtain Strong interaction confinement.

The difference between the Strong Interaction and the Higgs Mechanism was called the *Higgs – Confinement Dichotomy* in our earlier books.

5.2 Sublight vs. Superluminal Sectors in su(1,1) and F-Theory

F-Theory and su(1,1) which appears in the HyperCosmos and the Second Kind HyperCosmos have two time dimensions which lead to a more intricate form of sublight-superluminal separation. These separations lead to the splittings of symmetries in The Standard Model, the UST, and HyperCosmos spaces through the Dichotomy between the Higgs Mechanism and Strong Interaction Confinement. The su(1, 1) group appears as a subspace in all spaces $N \le 7$ above our $N = 7$ space.

This phenomena together with the structuring imposed by the HyperCosmos Theory almost completely specifies the separation of Internal Symmetry in the Standard Model and its extension in the UST and the HyperCosmos spaces.

5.3 Two Times Coordinates and Fermion Separations

The su(1, 1) group has a metric with two real time coordinates:

$$ds^2 = t_{01}{}^2 + t_{02}{}^2 - x_1{}^2 - x_2{}^2 \tag{5.7}$$

In Blaha (2007b) we showed that the four species of fermions: e-type, ν-type, q-up-type and q-down-type followed from the four types of boosts of the Lorentz Group $SO^+(1,3)$, which is often expressed as a representation of SL(2, **C**)). Complex Lorentz group boosts separate into sublight, superluminal, complex sublight, and complex superluminal boosts yielding the four fermion species respectively.

Now we have a more interesting situation with two time coordinates. Each has its own light speed. We take them to be numerically equal for the purposes of our discussion. One light speed divides the set of possible fermions into two "superspecies", namely Normal and Dark fermions. The second light speed further divides each superspecies into four parts. The result is the eight fermion species in our universe. It includes both Normal and Dark sectors. It is also the form of fermion species in the other HyperCosmos spaces which also have su(1, 1) multiple time coordinates.

The 16 fermions of an N = 7 Fundamental Reference Frame can be separated into the eight subspecies if account is taken of the occurrence of each quark species as a triplet. This separation of fermion species leads to the structure of the Unified SuperStandard Theory (UST) and the Standard Model. It also leads to the form of Fundamental Reference Frames for all HyperCosmos spaces.

A complete, detailed spectrum of fundamental fermions in our universe, and other universes, thus emerges, part of which appears in the Standard Model for our universe.

t_{01}:	sublight **Normal**				superluminal **Dark**			
t_{02}:	sublight		superluminal		sublight		superluminal	
	q-up-type	q-down-type	e-type	ν-type	q-up-type	q-down-type	e-type	ν-type
	6		2		6		2	

Figure 5.1. Separation of fundamental fermions. The superluminality of fermions may be masked by interactions as in the case of Higgs bosons.

5.4 Scalar Bosons of HyperCosmos Symmetries for N = 7

We now consider scalar bosons including Higgs bosons. This case is particularly interesting because it evidences the separation effects of the two time coordinates which lead to sublight and superluminal particle differentiation.

Unlike the fermion case the irreducible representations of the symmetry groups have complex dimensions that may be separated into real and imaginary parts. The FRF contains a set of 16 scalar bosons. These are first separated by time coordinate t_{01} into Normal and Dark bosons. Then t_{02} separates the bosons in each sector into Higgs and non-Higgs bosons mirroring the fermion case above. See Fig. 5.2.

In the Normal sector there is a non-Higgs set of three SU(3) bosons corresponding to the up-type and down-type quarks – totaling 6 bosons , and two Higgs bosons forming an SU(2) representation. The Dark sector has similar sets of bosons. See Fig. 9.2b.

t_{01}:	sublight **Normal** **U(4)** **8**		superluminal **Dark** **U(4)** **8**	
t_{02}:	sublight	superluminal	sublight	superluminal
	Non-Higgs	**Higgs Bosons**	**Non-Higgs**	**Higgs Bosons**
	6 SU(3)	**2** SU(2)	**6** SU(3)	**2** SU(2)

Figure 5.2. Bosons separated by su(1, 1)'s two times coordinates.

5.5 Internal Symmetries of the HyperCosmos for N = 7

This section shows the separation of internal symmetries due to the two time coordinates of su(1,1) – a group that appears in all HyperCosmos and all Second Kind HyperCosmos spaces.

t_{01}:	sublight	superluminal
	Normal	Dark
	U(4)	U(4)
Real-valued Representation Dimensions	8	8

t_{02}:	sublight	superluminal	sublight	superluminal
	6	2	6	2
	SU(3)	SU(2)	SU(3)	SU(2)

Figure 5.3. FRF Internal symmetry groups.

Note the sublight SU(3) group has a confining Strong interaction generated by a higher derivative interaction. Gravitation which should also be treated as sublight also has a higher derivative interaction that produces a MOND-like gravitational potential. See chapters 12 – 14 of Blaha (2022f) for details.

6. The Full HyperUnification Space

We now consider combined unification of all or some of the HyperCosmos spaces in a *Full HyperUnification Space* that will turn out to be a 42 space-time dimension space. Each space has a dimension array with d_{dN} elements. This array is reexpressed as a vector v_N with d_{dN} components residing in the space's unification space with space=time dimension r' given by eq. 3.1.

Now consider the sum of all ten HyperUnification spaces vectors:

$$v_S = \sum_{N=0}^{9} v_N \tag{6.1}$$

$$= 4^2 + 8^2 + 16^2 + 32^2 + 64^2 + 128^2 + 256^2 + 512^2 + 1024^2 + 2048^2$$

with

$$v_N = d_{dN} \tag{6.2}$$

These elements form a v_S-vector in the Full HyperUnification space. See Figs. 6.1 and 6.2. The sum of the ten vectors of the ten HyperUnification spaces vectors is a vector d_{cS} of the Full HyperUnification Space:

$$d_{cS} = v_S = 5{,}592{,}400 \tag{6.3}$$

There is a corresponding set of square dimension array blocks whose total is

$$d_{dS} = 4^4 + 8^4 + 16^4 + 32^4 + 64^4 + 128^4 + 256^4\text{^}+ 512^4 + 1024^4 + 2048^4$$
$$= 1.8765\times10^{13} \tag{6.4}$$

The Full HyperUnification Space plays the role of unification for all ten HyperCosmos spaces. Each of the ten HyperUnification space vectors of size d_{dN} in v_S individually has d_{dN} components. Correspondingly, each subvector has a $d_{dN} \times d_{dN}$ square array along the diagonal of a combined transformation which we call a *HyperUnification Transformation.* (Fig. 6.2)

Using eq. 3.1 we find the dimension array block defined by the d_{cS} vector is

$$d_H = 5{,}592{,}400^2 = 3.12749\times10^{13} \tag{6.5}$$

If we treat this square array d_H as corresponding to a HyperCosmos space then we can calculate the corresponding space-time dimension

$$r_H = \log_2(d_H/16) \tag{6.6}$$

$$= 2 \log_2(5{,}592{,}400) - 4 = 40.83$$

using

$$d_{dH} = 2^{22 - 2N_H} = 2^{r_H + 4} \qquad (6.6)$$

The dimension r_H is non-integral. Noting that space-time dimensions are even numbered in HyperCosmos spaces. we define the *Full HyperUnification Space* to have

$$r_H = 42 \qquad (6.7)$$

with

$$N_H = ½(18 - r_H) = -12$$

space-time dimensions.[31] It is the smallest space-time dimension that has a dimension array block containing the d_H dimension array block as shown in Fig. 6.2. (Note: the dimension array block d_{dN_H} has the same size as the General Relativistic transformation between the FRF and a static space-time.)

6.1 General Relativistic Transformation

The General Relativistic transformations of the $r_H = 42$ Full HyperUnification Space includes the General Relativistic transformations of any or all, HyperUnification spaces along its "diagonal" where blocks correspond to individual unification space General Relativistic transformations. See Fig. 6.1. It also includes much more in the form of arbitrary, not necessarily block diagonal, transformations that mix various HyperCosmos unification spaces.

The 42 space-time dimension array has the vector column size

$$d_{cN_H} = 2^{42/2+2} = 2^{23} = 8{,}388{,}608 \qquad (6.8a)$$

with the number of array elements being

$$d_{dN_H} = 8{,}388{,}608^2 = 7.03687 \times 10^{13} \qquad (6.8b)$$

Since d_{cN_H} is larger than d_{cS} there is an apparent excess in the 42 space-time dimension Full HyperUnification Space which we consider below. (See Fig. 6.2.)

We can define a set of HyperUnification Space General Relativistic transformations that transform in each of the individual HyperUnification sub-blocks separately. We can also define transformations that transform combinations of all 10 HyperUnification spaces including the entire set of HyperUnification spaces.

6.2 Full HyperUnification Space FRF

The FRF vector for the Full HyperUnification Space has d_{cN_H} components. Comparing it to the size of d_{cS} = 5,592,400-vector we find the General Relativistic

---

[31] Negative values of N are allowed to specify Cayley-number type dimension arrays. However they are not in the ProtoCosmos Model.

transformations of the 10 HyperCosmos unification spaces are a subset of the 42 space-time dimension HyperUnification Space transformations. The column excess is 2,796,208 dimensions. The ratio of the larger box labeled A to the smaller box labeled B in Fig. 1.2 is exactly

$$d_{cN_H}/d_{cS} = 1.5 = 3/2 \tag{6.9}$$

See Figs. 6.1 – 6.3 for the relations of the various parameters. Thus the B box column size d_{cB} satisfies

$$d_{cB} = ½\, d_{cS} \tag{6.10}$$

or

$$d_{cB} = ½ \sum_{N=0}^{9} v_N \tag{6.11}$$

Eq. 6.11 suggests that the excess of space within the 42 dimension space has a relevant interpretation. We explore it in chapter 8. It suggests that the excess—the B block— is a scaled down version of the A block. We call it the Full HyperUnification Space of the Second Kind. It corresponds to a new *HyperCosmos of the Second Kind* described in chapter 8.

The fact that the 42 dimension space directly contains what will be seen to be ***exactly*** *two HyperCosmos's supports the choice of a 42 dimension space.*

We will use HyperUnification Space transformations in the next chapter to reduce the set of all HyperCosmos spaces'dimensions to one universal, primordial dimension.

We note that the ratio of the square dimension arrays labeled A and B in Fig. 6.2 is exactly

$$d_{dN_H}/d_{dS} = 7.03687 \;{*}10^{13}/\, 1.8765E{+}13 \tag{6.12}$$
$$= 3.75 = 15/4$$

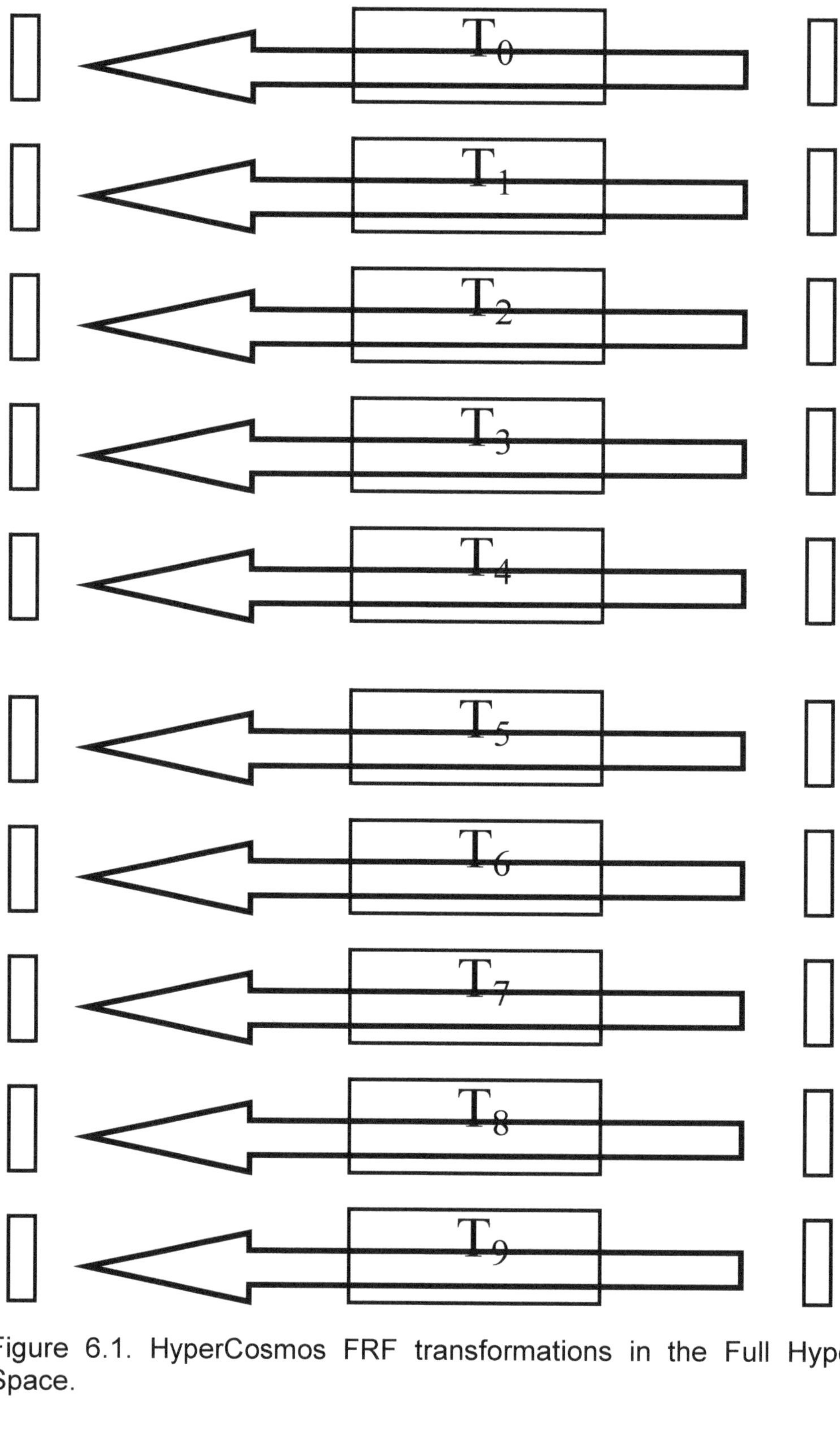

Figure 6.1. HyperCosmos FRF transformations in the Full HyperUnification Space.

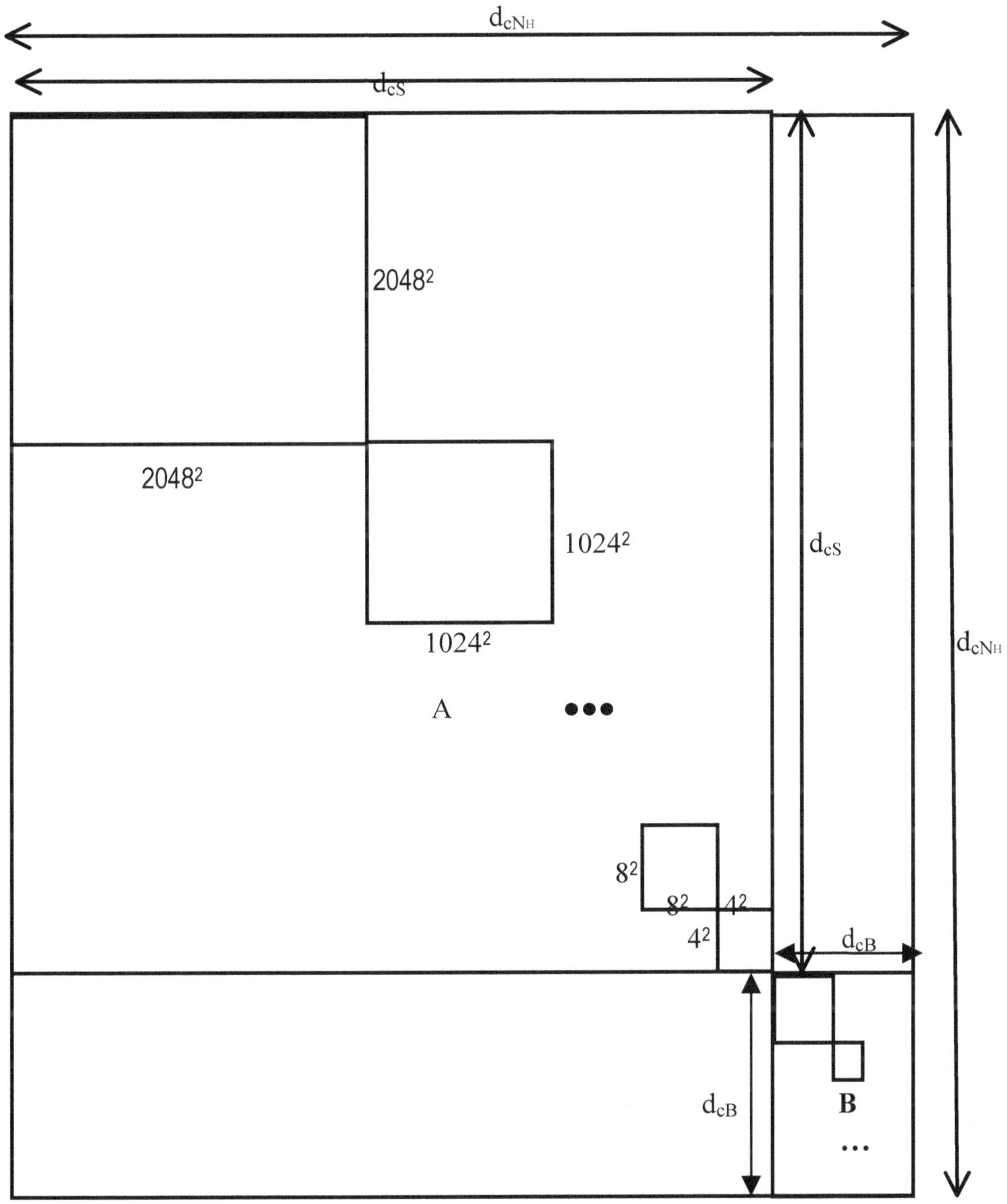

Figure 6.2. Form of a HyperUnification transformation. It is also the form of the dimension array of 42 dimension space-time. All blocks are square. The figure is not drawn to scale.

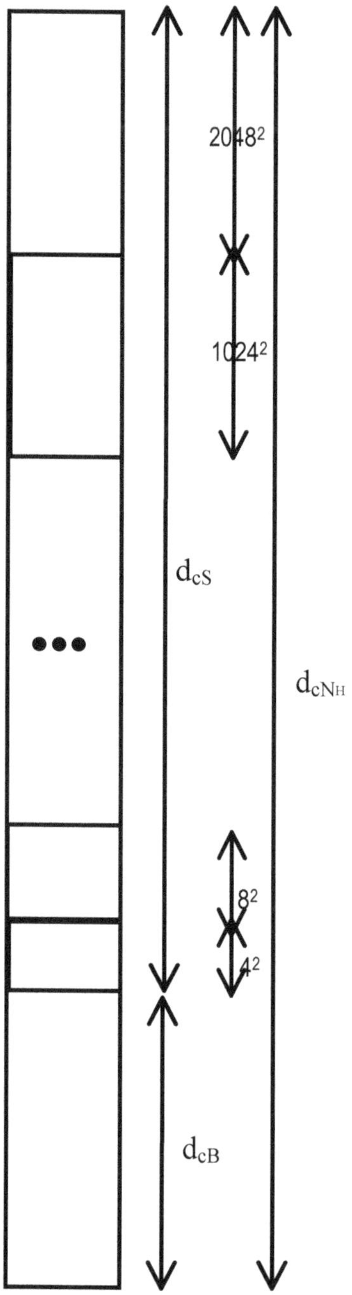

Figure 6.3. The r = 42 space-time HyperUnification vector form.

7. One Dimension Generation in the Full HyperUnification Space and FRF

The FRF of a single HyperUnification space was reduced to one dimension in section 4.4. It is possible to reduce the combined FRF of the Full HyperUnification Space of Fig. 7.1 to one dimension.

We demonstrate this possibility with a simple example. We reassemble the transformation to the form of Fig. 7.1 placing all space blocks of both the HyperCosmos and the Second Kind HyperCosmos to the extreme left.

Assuming the FRF has one non-zero dimension (element) a in row 1 of the Full HyperUnification Space FRF vector with all other components zero we define the vector

$$V = (a, 0, 0, \ldots) \tag{7.1}$$

and a *prototype* 42 dimension General Relativistic transformation

$$[T]_{ij} = b_i \ \delta_{i1} + (1 - \delta_{1i})(1 - \delta_{j1})c_{ij} \tag{7.2}$$

where $b_i \neq 0$ and the c_{ij} are numeric. Then T has entries b_i in column 1 and a matrix c_{ij} in the rows and columns for i, j = 2, 3, … . Consequently a full non-zero set of components d_{cN_H} is generated in the output vector which map to the d_{cN_H} dimensions in vector V'. It is generated in a vector from one FRF dimension (element):

$$V' = TV \tag{7.3}$$

Thus we can generate the complete set of HyperCosmos and/or Second Kind HyperCosmos dimension arrays from a one dimension element in the 42 space-time dimension's FRF. The vector has non-zero components which become d_{cN_H} dimensions.

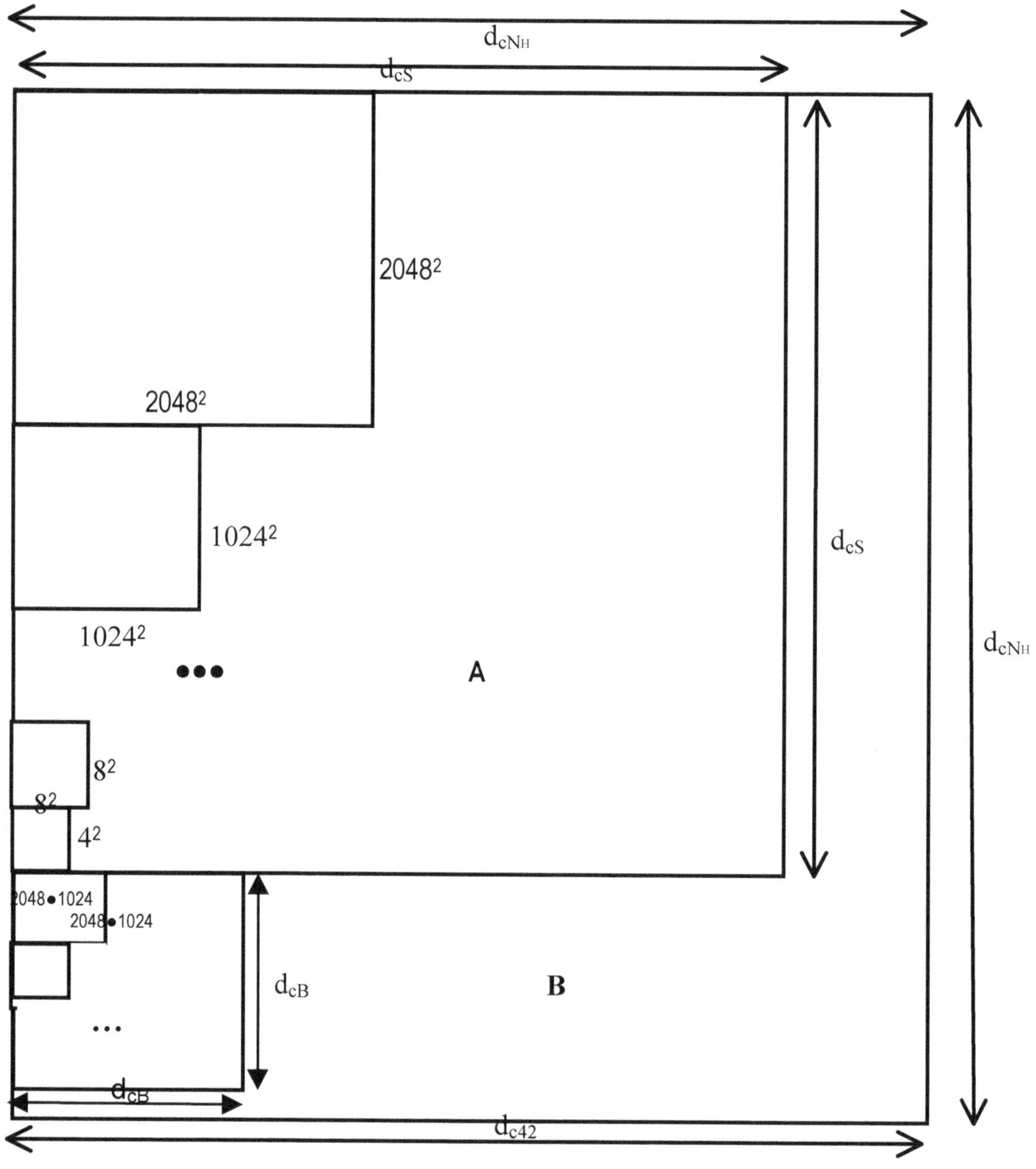

Figure 7.1. A HyperUnification transformation restructured to generate all dimentions from a one dimension FRF. The diagonal blocks are shifted to the left to support the transformation from a one dimension FRF of the 42 space-time dimension space. All blocks are square. This figure is not drawn to scale.

8. The HyperCosmos of the Second Kind

The form of d_{cR} for array B in chapter 6 suggests that the HyperCosmos of the Second Kind spectrum of spaces is similar to our original HyperCosmos with one basic change in the numerics:

$$d_{cR} = \frac{1}{2} \sum_{N=0}^{9} v_N \tag{8/1}$$

The vector v_N

$$v_N = d_{dN} \tag{8.2}$$

should be modified to

$$v_N' = \frac{1}{2} d_{dN} = d_{dN}' \tag{8.3}$$

which suggests

$$d_{dN}' = \frac{1}{2} 2^{r/2+2} 2^{r/2+2} = 2^{r/2+1} 2^{r/2+2} = 2^{r+3} \tag{8.4}$$

The Second Kind arrays are not square. They are rectangular.

There are several ways to obtain a $2^{r/2+1}$ factor:

1. Change the internal symmetry factor by a factor of two. This can be accomplished by eliminating the Dark sector: no Dark fermions and no Dark interactions. This would explain the existence of Dark matter without any interactions except gravity.

2. Remove the PseudoQuantum quantum field framework. This approach would remove the advantages of PseudoQuantum field theory.

3. Restrict General Relativistic (GR) transformations to b an $b^\dagger$ operators only. The d and $b^\dagger$ operators would not undergo GR transformations. These operators are half of the vector operators in GR transformations as evident in eq. 3.3 in chapter 3.

It appears that choice 1 is the best since it conforms to the known Dark matter features and best preserves a HyperCosmos formalism.

Thus a HyperCosmos of the Second Kind is possible. Whether Nature chooses it is an open question. We shall continue using the original HyperCosmos formalism.

The 42 space-time dimension space has a General Relativity that can mix the HyperCosmos spaces and the Second Kind HyperCosmos spaces. Thus there are gravitational interactions between them if the Second Kind set of spaces is instantiated with universes. We regard the HyperCosmos and the Second Kind HyperCosmos as siblings with the only interaction between them being gravitation at best.

The particles and symmetry groups of the Second Kind HyperCosmos spaces are not the same as the particles and symmetry groups in HyperCosmos spaces. The following figures show features of the N = 7 Second Kind HyperCosmos space and thus a Second Kind QUeST and Second Kind UST.

8.1 Two HyperCosmoses

We see that there are two HyperCosmoses: the original HyperCosmos and the HyperCosmos of the Second Kind. Either or both may be instantiated with universes. Our universe, at the moment, may be of either HyperCosmos. The key undecided question is the existence of a Dark sector.

The form of the Full HyperUnification Space supports two HyperCosmoses exactly. It provides significant support for the choice of a 42 dimension space-time.

Whether our universe is in the original HyperCosmos or in the Second Kind HyperCosmos is not known at present. The choice depends on the existence of the Dark sector.

8.2 Types of Transformations

The 42 space-time dimension Full HyperUnification space General Relativistic transformations are:

1. Transformations on the FRF of each HyperCosmos HyperUnification space individually.

2. Transformations on the FRF of each Second Kind HyperCosmos HyperUnification space individually.

3. Combined transformations on the FRFs of HyperCosmos HyperUnification spaces.

4. Combined transformations on the FRFs of Second Kind HyperCosmos HyperUnification spaces.

5. Transformations on the complete set of Full HyperUnification Space FRFs of both HyperCosmos and Second Kind HyperCosmos HyperUnification spaces.

8.3 Second Kind HyperUnification Spaces

Each of the ten Second Kind HyperCosmos spaces has a corresponding HyperUnification space. We assume the set of space-time dimensions of the Second Kind HyperCosmos is the same as the HyperCosmos. For space-time dimension r there are 2^{r+3} by eq. 8.4. Consequently the corresponding Second Kind HyperUnification space has vectors with 2^{r+3} components as indicated in Figs. 6.2 and 7.1.

Since the relation of the space-time dimensions is

$$2^{r+3} = 2^{r'/2+2} \tag{8.5}$$

for HyperCosmos-like spaces we find

$$r' = 2r + 2 \tag{8.6}$$

where r' is the space-time dimension of the HyperUnification space.

HYPERCOSMOS OF THE SECOND KIND SPACES SPECTRUM

Blaha Space Number $N = O_s$	Cayley-Dickson Number n	Cayley Number d_c	Dimension Array size d_d	Space-time-Dimension r	CASe Group $su(2^{r/2},2^{r/2})$ CASe
0	10	1024	1024 × 2048	18	su(512,512)
1	9	512	512 × 1024	16	su(256,256)
2	8	256	256 × 512	14	su(128,128)
3	7	128	128 × 256	12	su(64,64)
4	6	64	64 × 128	10	su(32,32)
5	5	32	32 × 64	8	su(16,16)
6	4	16	16 × 32	6	su(8,8)
7	**3**	**8**	**8 × 16**	**4**	**su(4,4)**
8	**2**	4	4 × 8	2	su(2,2)
9	**1**	2	2 × 4	0	su(1,1)

Figure 8.1. The HyperCosmos of the Second Kind space spectrum. The space for our universe, is number 7, with Cayley-Dickson number 3 (which corresponds to octonions) is in bold type. Note changed d_d column relative to the HyperCosmos.

Number of Columns =	4 NORMAL	4	4 DARK	4
Layer 1 4	e 3 up-quarks	ν 3 down-quarks	e 3 up-quarks	ν 3 down-quarks
Layer 2 4	e 3 up-quarks	ν 3 down-quarks	e 3 up-quarks	ν 3 down-quarks
Layer 3 4	e 3 up-quarks	ν 3 down-quarks	e 3 up-quarks	ν 3 down-quarks
Layer 4 4	e 3 up-quarks	ν 3 down-quarks	e 3 up-quarks	ν 3 down-quarks

Figure 8.2. The HyperCosmos of the Second Kind 8×16 fermion spectrum tentatively arranged as SU(4)-plets that correspond directly with SU(4) (or SU(3)⊗U(1)) fermions. There are four layers. Each set of 4 fermions has 4 generations matching the number of rows in each layer. This Periodic Table is broken into Normal and Dark sectors. No Dark sector as indicated by the darkened part.

	NORMAL	DARK
Layer 1	U(4)⊗U(4) U(4)⊗U(4)	U(4)⊗U(4) U(4)⊗U(4)
Layer 2	U(4)⊗U(4) U(4)⊗U(4)	U(4)⊗U(4) U(4)⊗U(4)
Layer 3	U(4)⊗U(4) U(4)⊗U(4)	U(4)⊗U(4) U(4)⊗U(4)
Layer 4	U(4)⊗U(4) U(4)⊗U(4)	U(4)⊗U(4) U(4)⊗U(4)

Figure 8.3. The Second Kind "initial" distribution of sets of N = 7 symmetry groups. Each set is distinct and supports interactions only for the corresponding set of fermions (separately for Normal and Dark fermions). *Thus each set of 16 fermion generations has its own quantum numbers and interactions.* Each U(4)⊗U(4) set has a 16 real-valued dimension representation, which are importance when we consider Fundamental Reference Frames. There is no Dark sector as indicated by the darkened part.

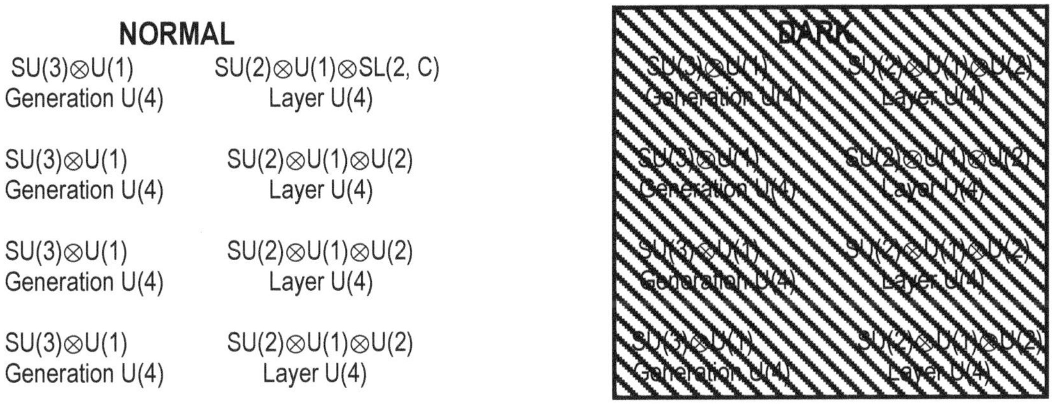

Figure 8.4. The transformed/broken sets of symmetries in N = 7 n Second Kind HyperCosmos space. Note each element has a 16 real dimension representation. This depiction is also evident in the Second Kind QUeST and UST. The SL(2, **C**) representation has four coordinates.[32] There is no Dark sector as indicated by the darkened part.

---

[32] The Lorentz Group $SO^+(1, 3)$ is often specified with an SL(2, **C**) representation.

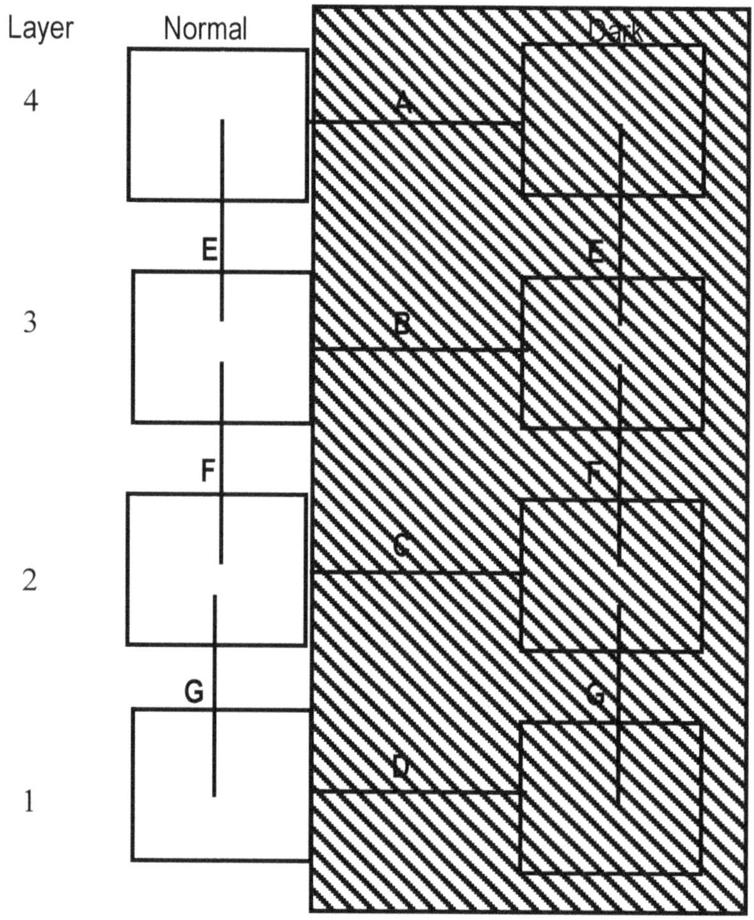

Figure 8.5. The three U(2) Connection groups[33] (shown as 3 lines) between the eight QUeST/UST blocks in the N = 7 Second Kind HyperCosmos. The Darkened part is not present in the Second Kind case. Connection groups are obtained by transfering 12 dimensions from QUeST space-time to internal symmetries with the consequent reduction of the space-time from four octonion (complex quaternion) coordinates to four real coordinates. The Connection groups generate rotations and interactions between corresponding fermions and vector bosons of each pair of blocks. This figure is abstracted from the corresponding figure in Appensix 10-B.

[33] Connection groups are discussed in Appendix 9-B.

9. The UltraUnification (UU) Space of the Full HyperUnification Space

The Full HyperUnification Space is a 42 space-time dimension space. It has a set of transformations of the form of Fig. 6.2. It also has a square dimension array of the same form as the transformation array: The number of rows and columns is d_{c42}. The dimension array has similar content of dimension arrays of HyperUnification spaces of the HyperCosmos and the Second Kind HyperCosmos corresponding to the blocks of Fig. 6.2.

We can specify a set of unified transformations of these blocks in the same manner as we originally did for HyperCosmos spaces when we defined HyperUnification spaces using

$$r' = 2r + 4 \tag{4.8}$$

with r = 42. The space-time dimension of its HyperUnification space, which we call the *UltraUnification (UU) Space*, is r' = 88. A 88 dimension General Relativistic transformation generally transforms all the contents of the Full HyperUnification Space. It therefore unifies all the space-time and Internal Symmetries within the set of HyperCosmos spaces and the Second Kind HyperCosmos spaces.

The UU space is needed to have an FRF with one non-zero dimension be transformed to the full set of dimensions in the 42 space-time dimension Full HyperUnification space and thence to the complete set of dimensions in the ten HyperCosmos spaces and in the ten Second Kind HyperCosmos spaces.

Note:

The defining feature of a HyperUnification space with space-time dimension r' of a space with space-time dimension r is that the number of components in the dimension r' GR transformation vector equals the number of components d_{dN} *in the dimension array of the space of space-time dimension r.*

The 88 dimension UltraUnification space completes the Cosmos.[34] *The Cosmos has four levels as shown in Fig. 9.1. The total number of Cosmos spaces is*

!0 HyperCosmos spaces + their 10 HyperUnification spaces + (9.1)
+ !0 Second Kind HyperCosmos spaces +

---

[34] The appearance of the dimensionless integers 42 and 88 in this Cosmos evokes the memory of the Eddington and Dirac (and others) conjectures on the appearance of 40 and 80 approximately in various combinations of universe parameters and physical constants. If some of their conjectures are valid they would be evidence for the author's Cosmos Theory. The dimensionless numbers 40 and 80 share the property of dimensionlessness with the dimensions of Cosmos Theory.

+their 10 Second Kind HyperUnification spaces +
+ the 42 Dimension Full HyperUnification Space +
+ the 88 Dimension UU space = 42 spaces

We thus find the interesting equality of the Full HyperUnification space dimension 42 and the total number of Cosmos spaces. Whether this equality reflects a deeper underlying relation within the Cosmos' structure remains to be determined.[35] Fig. 9.1 displays the four levels and the 42 spaces of the complete author's *Cosmos Theory*.

The four levels of Fig. 9.1 may be viewed as forming a tetragram that reflects the structure of the Physical Reality (being) of the Cosmos. The Cosmos may also be a reflection of the Prime Mover – The *Tetragrammaton* – based on the 42 dimension Full HyperUnification space, the 42 Cosmos spaces, and the fourfold structure of the Cosmos.

9.1 Relation Between the Four Levels

The four levels of the Cosmos (Fig. 9.1) are intimately related:

A. The fourth level consists of the ten HyperCosmos spaces augmented by ten Second Kind HyperCosmos spaces.

B. The third level consists of 20 HyperUnification spaces. They are the HyperUnification spaces of the HyperCosmos and of the Second Kind HyperCosmos. The fourth level spaces each have a second level HyperUnification space. General Relativistic transformations of the HyperUnification space generate the dimension array of each r dimension HyperCosmos and Second Kind HyperCosmos space from a vector in the space's FRF. The vector may have only one non-zero dimension or may have a set of $2^{r/2+2}$ non-zero dimensions (or some other number of non-zero dimensions).

C. The second level is a 42 space-time dimension space with the same overall form as a HyperCosmos space. In this space the 20 HyperUnification spaces of the third level appear "along the diagonal."

D. The first level is the 88 space-time dimension UltraUnification space of the third level 42 dimension space. General Relativistic transformations of the UltraUnification space generate vectors yielding the dimension array of the 42 dimension third level HyperCosmos space from a vector in the fourth level space's FRF. The vector may have only one non-zero dimension or a set of $2^{42+4} = 2^{46}$ non-zero dimensions (or some other number of non-zero dimensions).

---

[35] The number 42 has a significance in a number in religious settings. We will not consider it in that context here.

A General Relativistic transformation of the 88 space-time dimension UltraUnification space cascades down to the fourth level HyperCosmos and Second HyperCosmos dimension arrays. Consequently it is possible to populate all HyperCosmos and Second Kind HyperCosmos dimension arrays with one non-zero dimension in the FRF vector of the 88 space-time dimension UltraUnification space.

9.2 Reduction of the UltraUnification Space FRF to One Dimension

We demonstrate this possibility with a simple example. Assuming the UU FRF has one non-zero dimension (element) a in row 1 of the UU FRF vector with all other components zero we define the vector with d_{c88} components as

$$V = (a, 0, 0, \ldots) \tag{9,2}$$

and a *prototype* 88 space-time dimension General Relativistic transformation

$$[T]_{ij} = b_i \ \delta_{i1} + (1 - \delta_{1i})(1 - \delta_{j1})c_{ij} \tag{9.3}$$

where $b_i \neq 0$ and the c_{ij} are numeric. Then T has entries b_i in column 1 of all rows and a matrix c_{ij} in the rows and columns for i, j = 2, 3, … . Consequently a vector

$$V' = TV \tag{9.4}$$

is generated with all components non-zero. The generated vector V' has $d_{cN} = 2^{46}$ non-zero components which become d_{cN} dimensions. This vector can be written as a square array with 2^{46} components[36] where Blaha number[37] N = –35 using

$$N = 9 - r/2 \tag{9.5}$$

are generated in a vector from one FRF dimension (element). Having the 42 space-time dimension array one can then map to the individual HyperCosmos and Second HyperCosmos dimension arrays.

We can generate the complete set of HyperCosmos and/or Second Kind HyperCosmos dimension arrays from a one dimension element.

---

[36] In agreement with eq. 6.8a.

[37] Negative values of N are allowed. They do not appear in the ProtoCosmos model.

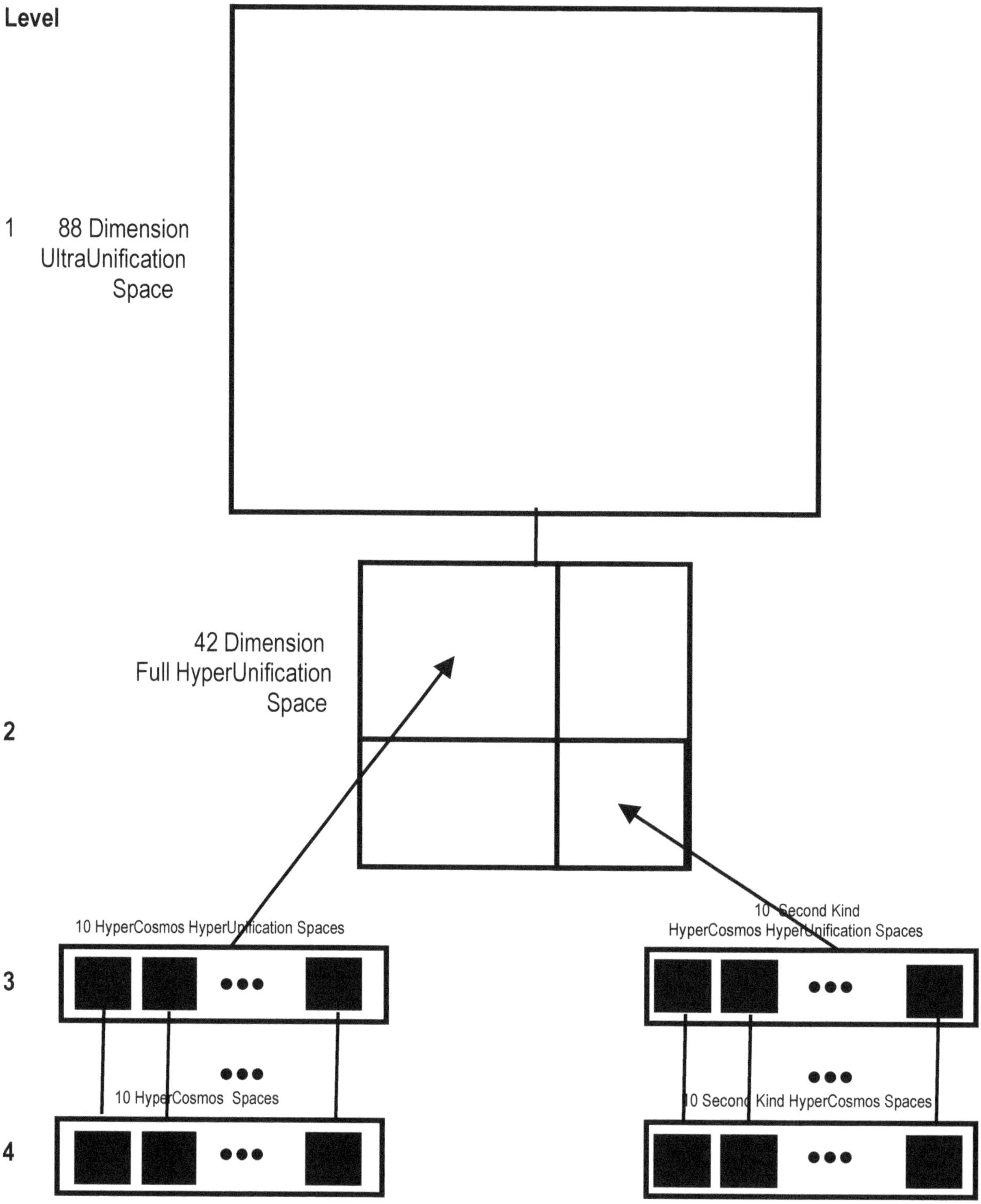

Figure 9.1. Diagram of the four levels of the Cosmos. They contain 42 spaces.

10. The Generation of the Cosmos

Ideally the Cosmos may be viewed as created from nothing and then evolving according to the dictates of "inexorable" Physical Law. Our Cosmos almost meets that goal. The Prime Mover starts from one dimension—a primitive term, undefined except that it has no value or substance—treatable as a "mental" concept. The dimension exists in the Fundamental Reference Frame (FRF) of a space with 88 space-time dimensions whose dimensions simply specify coordinates.

Treating this space as the HyperUnification of a 42 space-time dimension space we define a set of 88 space-time dimension General Relativistic (GR) transformations that take a vector of d_{cN_H} components (the number of elements in the 42 dimension array of dimensions; See eq. 6.8a.) with only one non-zero component (the original dimension) and generate a vector with d_{cN_H} non-zero components (dimensions).

One might ask if the transformation creates dimensions. In a physical sense the answer is no. The GR transformation "merely" relates the "situation" in the FRF to the "situation" in the destination reference frame. In the 1970s it became clear that the particles seen in one reference frame are equivalent to a different set of particles in an accelerating reference frame. No particles were created. Reference frames determine what one sees in general.

The above process does have a result: the conceptual expansion of the one original dimension to the dimensions of the second, third and fourth levels of Fig. 9.1. And so the Prime Mover has generates, without creating, the framework of the Cosmos.

Having populated the 42 dimension Full HyperUnification space dimension array, (chapters 6 and 7) we see that each HyperCosmos space and each Second Kind HyperCosmos space has its HyperUnification space dimension array populated. The dimensions in each HyperUnification space dimension array may be used to define the dimension array of the corresponding HyperCosmos and/or Second kind HyperCosmos space.

Up to this point nothing has been created. All dimensions flow from the original one dimension in the 88 dimension UU space. The structure of the Cosmos has been *conceptually* generated from one original dimension. And that is all that is needed to support Creation.

10.1 Structure of the Cosmos

The structure of the UU space and the spaces on levels 2 – 4 determines the overall structure of the HyperCosmos and Second Kind HyperCosmos spaces. The FRF contents are specified in chapters 4 and 5 by considering the forms of symmetry groups and fermions implied by the two time coordinates generated through su(1,1): a group that appears in all spaces. After a General Relativistic transformation in each space's HyperUnification space the dimension array of each space emerges. The dimension array of each HyperCosmos and Second Kind HyperCosmos space consists of replicates of its FRF content. The pattern (splittings) of internal symmetries of each space is thereby determined. The fundamental fermions of each space are similarly determined as are vector gauge particles and scalar Higgs and non-Higgs bosons.

Thus the four levels of spaces in Fig. 9.1 fully determine the symmetry groups and particles associated with each space.

10.2 Creation of Universes – ProtoCosmos

Given the definition of spaces and their contents derived above we can turn to the creation of universes.[38] The UniDimension ProtoCosmos Model was shown in Blaha (2022g) to be derivable from one dimension where the dimension is fractionated[39] into an infinite set[40] of "dust." The dust which is formless may be shaped into a line segment [-1, 1] and transformed to a continuous line. A dynamics may be introduced consisting of a one-dimension Dirac equation with a dimensionless 1/x potential. The spectrum of this equation can be shaped by setting constants to produce a set of eigenfunctions: each with a fermion wave function factor. Thus each HyperCosmos space[41] can be instantiated by creating states. These states specify universes that begin with a certain energy and dimension array.

The ProtoCosmos generates the spaces and universes of the HyperCosmos from one dimension. By making the ProtoCosmos energy spectrum consistent with the HyperCosmos (and/or Second Kind HyperCosmos) spectrum we can *consistently* specify the one dimension of the UU FRF as the *source* of the ProtoCosmos one dimension in order to create a full Cosmos.

The result is a Cosmos populated with spaces and universes containing symmetries and matter. By universes we mean instances of spaces. This definition includes our universe, possible Multiverses, and universes containing universe.

The initial creation of universes seems to require the specification of universe states using the ProtoCosmos wave function formalism. Subsequently, if universes have interactions and a Quantum Field Theory formulation[42] they may generate additional universes through collisions. They may also generate subuniverses. The author's previous books describe these possibilities.

10.3 Evolution in Each Universe

When a universe is created in an enclosing universe it appears likely that it was created with a very large amount of internal energy at a point. Thermodynamics tells us that it will expand. This author developed a Big Bang theory describing the expansion process and the resultant Hubble expansion.[43]

---

[38] We take universes to mean an instantiation of a space. A universe may be our universe, a multiverse, or a set of interrelated universe within a host space.

[39] See Blaha (2022d).

[40] The reduction of a dimension to dust allows the Prime Mover to create the Cosmos from the "nearest thing" to Nothing.

[41] Or alternately for each Second Kind HyperCosmos space.

[42] See Blaha (2021f) and (2022c) for example.

[43] See Blaha (2004) and (2021d).

There is no need for a Prime Mover after universes (and the Cosmos of Fig. 9.1) appear. The universes evolve by the strict laws of dynamics generated from the symmetry group forms of the spaces and the particles specified from the FRFs of the HyperCosmos spaces.

10.4 Dynamical Laws of a Universe

The dynamical laws of a universe are specified by:

1. Dirac equations for fermions; and Yang-Mills equations for vector bosons with the details of the equations are determined by the relevant symmetry groups. General Relativistic generalizations of the equations are immediate. Scalar boson particles are also determined from the FRF of the universe's HyperCosmos space. See chapters 4 and 5. See chapters 12 – 14 of Blaha (2022f) for the PseudoQuantum formulation of the Riemann-Christoffel curvature tensor whose terms lead to the Euler-Lagrange formulation of the dynamical equations.

2. A PseudoQuantum formulation of the wave functions, states, and dynamical equations.

10.5 Conclusion

We conclude from the above discussion that all is based on dimension—the one primordial dimension resident in the FRF of the UltraUnificationspace.

11. QUeST/UST Structure

Over the past years we developed the Unified SuperStandard Theory (UST). In this chapter we outline its development and then proceed to QUeST, which implements the UST within the number N = 7 HyperCosmos number space.

The HyperCosmos structure for all spaces completely defines the fermions and the splittings of Internal Symmetry groups of the spaces. The dynamics of the UST universe are determined by the Quantum Theory and Lagrangian field equations.

11.1 The Standard Model of Elementary Particles

The Standard Model is known to have an SU(3)⊗SU(2)⊗U(1) internal symmetry. It has three known generations of fermions.

11.2 UST

The UST developed over a number of years had the Standard Model embedded in it. It had a more complex structure with four generations of fermions in both a Normal sector and a Dark sector.

This author introduced two additional U(4) internal symmetry groups: the Generation group and the Layer groups. The Generation group caused the fermion spectrum to be enlarged to four generations. The Layer groups caused the four generations of fermions to become four layers. See Appendix 11-A for details. A Connection group was found necessary. Its role was to "unite" the various parts of the fermion spectrum. See appendix 11-B for details.

The total number of generations and layers became 16. A Normal and Dark sector of fermions appeared in the UST.

11.3 QUeST

In the Fall, 2019 this author found the UST could be generalized to the form of QUeST space. QUeST had the Standard Model embedded in it. It had a more complex structure with the UST four generations of fermions in both a Normal sector and a Dark sector. Each of the four layers of fermions and an independent set of internal symmetries.

In the Fall of 2019 the author determined that the UST and QUeST had the same form. The equivalent forms led the author to proceed to develop the HyperCosmos spectrum of spaces.

11.4 HyperCosmos Spaces

In the past three years the author found that the N = 7 space of our universe was but one of ten spaces. The other HyperCosmos spaces had similar forms in higher space-time dimensions.

Recently the Cosmos of Fig. 9.1 was developed containing 42 spaces. The HyperCosmos and Second Kind HyperCosmos are embedded within the Cosmos together with unification spaces.

Appendix 11-A. Generation and Layer Groups

11-A.1 U(4) Generation Groups

In the Big Bang all particles were massless and all symmetries unbroken. Hence the four Normal particle number symmetries, and the four Dark particle number symmetries, are all "conserved" in the Big Bang. Afterwards conservation laws are then broken.

We define two particle number operators for normal up-quark particles and down-quark particles, B_{uq} and B_{dq}. Similarly we define two particle number operators for normal species "e" (electron) particles and species "ν" particles, B_e and $B_ν$. Similarly we define Dark matter equivalents:[44] B_{De}, $B_{Dν}$, B_{Duq}, and B_{Ddq}.

In the absence of symmetry breaking these fermion particle number operators would be conserved. Thus there are two sets of "diagonal" operators with associated U(4) groups for the Normal and Dark sectors. They are part of the Normal U(4) Generation Group and the Dark U(4) Generation Group.

The fermion fundamental representation of a U(4) group has four fermions. U(4) has rotations, and also interactions of the form $\overline{\Psi}\gamma\cdot B\cdot T\Psi$ where Ψ is a fermion four-vector, B is a 16 component U(4) gauge field, and T consists of 16 component 4 × 4 U(4) arrays.

In the case of the Generation Group the gauge fields have electric charge zero. Since the four species have different electric charges (1, 0, 2/3, -1/3) the U(4) gauge boson fields cannot mix the fermions of different species. Generation Group interactions are diagonal[45] in fermion species (e, ν, up-quark, and down-quark species).

Consequently the U(4) Generation Group for each layer[46] must have a reducible representation D consisting of a set of four fundamental U(4) representations, D_e, $D_ν$, D_{upq}, and D_{dnq}, appearing in blocks along the diagonal of D. Each block is a separate U(4) irreducible representation for a species (due to the electric charge superselection rule.) There is a U(4) Generation group for each of the four layers of the Normal and Dark sectors totaling to 8 generation groups.

There are four generations of each species in the Normal and in the Dark matter sectors. The four generations for each fermion species: e, ν, up-quark, and down-quark each furnish a U(4) fundamental representation within the reducible representation D. The fourth generation of normal fermions has not as yet been found due to their extremely large masses.

---

[44] By analogy, we assume that there are four species of Dark matter: charged Dark leptons, neutral Dark leptons, Dark up-type quarks, and Dark down-type quarks. Thus we are led to the Dark particle numbers: Dark Baryon Numbers, and Dark Lepton Numbers shown above.

[45] ElectroWeak interactions can cross between species due to their charged gauge vector bosons.

[46] For the Normal and Dark sectors separately.

The Generation Group rotates the fundamental fermions of each fundamental representation separately for each of the four species of each of the four layers.[47] Thus the Generation Group guarantees that all generations of each species have the same electric charge and other quantum numbers.

The U(4) Generation Group also specifies a gauge field interaction among the fermions of its fundamental representation, species by species, for both Normal and Dark sectors. The form of the interactions for the Normal sector for each fermion layer is:

$$g_e\Psi_e\gamma\cdot B_e\cdot T\Psi_e + g_\nu\bar{\Psi}_\nu\gamma\cdot B_\nu\cdot T\Psi_\nu + g_{upq}\bar{\Psi}_{upq}\gamma\cdot B_{upq}\cdot T\Psi_{upq} + g_{dnq}\bar{\Psi}_{dnq}\gamma\cdot B_{dnq}\cdot T\Psi_{dnq} \qquad (11\text{-}A.1)$$

where g_e ... are coupling constants, the gauge vector fields are B_e ... , and the Ψ_e ... are 4-vectors of fermions of the four generations of each species in a layer.

The gauge vector bosons of the Generation Group have large masses. If the conservation of the fermion particle numbers is broken then we view it as a consequence of Generation Group symmetry breaking.

Generation Group rotations guarantee the internal quantum numbers of each generation of each species are the same since symmetry breakdown is not present at the instant of the Big Bang.

The above discussion applies similarly to the Dark sector. Thus there are 8 Generation Groups in total.

11-A.2 U(4) Layer Groups

The set[48] of particle number operators can be extended if we take account of the fourfold fermion generations.

We can subdivide the above particle number sets into four additional particle numbers *per generation.* For the ith generation (of the four generations) we define

L_{ie} – The "e" species particle number for the ith generation
$L_{i\nu}$ – The ν species particle number for the ith generation
L_{iuq} – The up-quark species particle number for the ith generation
L_{idq} – The down-quark species particle number for the ith generation

L_{iDe} – The Dark "e" species particle number for the ith generation
$L_{iD\nu}$ – The Dark ν species particle number for the ith generation
L_{iDuq} – The Dark up-quark species particle number for the ith generation
L_{iDdq} – Dark down-quark species particle number for the ith generation

for each generation i = 1, 2, 3, 4. Individual fermions have positive $L_{ia} = +1$ values and antifermions have negative $L_{ia} = -1$ values for each species.

---

[47] There are separate Generation groups for each layer.
[48] Here again, in the Big Bang all particles were massless and all symmetries unbroken. Hence particle numbers are "conserved" in the Big Bang. Conservation is then broken afterwards in most cases.

At this point we have a set of four particle number operators for each of four generations (i = 1, 2, 3, 4) of fermions in the Normal sector and similarly in the Dark sector. We then define a U(4) group framework for each set of particle numbers.

The only way to specify fundamental representations for each of the four sets in a sector is to assume there are four layers, with each layer having four generations, and with a fundamental U(4) representation defined for each generation composed of fermions from each layer. Thus there are four Layer Groups for each Normal and each Dark sector: a Layer Group for generation 1, a Layer Group for generation 2, and so on.

The Layer Groups are also "split" by species due to the electric charge superselection rule. Each Layer Group is diagonal in the four fermion species. All their gauge fields are electrically neutral. Thus the Layer Group fundamental representations of the Normal sector total 16.[49] The Layer Group fundamental representations of the Dark sector also total 16. There are 4 Normal Layer Groups and 4 Dark Layer Groups.

Consequently each of the four U(4) Layer Groups in the Normal fermion sector has a reducible U(4) representation D_j for j = 1, 2, 3, 4. Each reducible representation is composed of four irreducible U(4) representations for each species due to the electric charge superselection rule:

$$D_j = D_{je} + D_{jv} + D_{jupq} + D_{jdnq},$$

for j = 1, 2, 3, 4.

There are four layers of each species in the Normal and in the Dark matter sectors. The second, third and fourth layers of normal fermions has not as yet been found due to their extremely large masses.

A Layer Group rotates the fundamental fermions of each fundamental representation separately for each of the four species of each of the four generations.

The Layer Groups guarantee that all layers of each species have the same electric charge and other quantum numbers.

Each U(4) Layer Group also specifies a gauge field interaction among the fermions of its fundamental representation, species by species, for both Normal and Dark sectors. The form of the interactions is:

$$g_{ei}\overline{\Psi}_{ei}\gamma\cdot C_{ei}\cdot T\Psi_{ei} + g_{vi}\overline{\Psi}_{vi}\gamma\cdot C_{vi}\cdot T\Psi_{vi} + g_{upqi}\overline{\Psi}_{upqi}\gamma\cdot C_{upqi}\cdot T\Psi_{upqi} + g_{dnqi}\overline{\Psi}_{dnqi}\gamma\cdot C_{dnqi}\cdot T\Psi_{dnqi} \tag{11-A.2}$$

for i = generation = 1, … , 4, where g_{ei} … are coupling constants, the gauge fields are C_{ei} … , and the Ψ_{ei} … are 4-vectors of fermions formed of the ith generation fermions in each layer of each species.

The gauge vector bosons of the Layer Groups also have large masses. If the conservation of the fermion particle numbers is broken then we view it as a consequence of Layer Groups symmetry breaking.

---

[49] Four Layers groups irreducible representations for the generations × four species = 16 irreducible representations.

Layer Group rotations guarantee the internal quantum numbers of each layer of each species are the same since symmetry breakdown is not present at the instant of the Big Bang.

The above discussion applies similarly to the Dark sector. There are 16 Dark Layer Groups.

Fig. 7.3 shows the fundamental fermion spectrum with the representations of the Generation groups and Layer groups indicated.

Experimentally, we know of three generations of fermions—the lowest 3 generations of the lowest level. The remaining 4th generation and three layers of fermions are of much higher mass and are yet to be found.

See Blaha (2019g) and (2018e) for a detailed discussion of the Layer Groups. We note in passing that the symmetries of these number operators are badly broken. Yet the underlying group structure remains.

11-B. Connection Group Symmetries and HyperCosmos Space-Time Coordinates

Hypercomplex numbers are known to be related to symmetry groups. In this section we consider the multiple space-time symmetries that appear in the separation of dimension arrays into representations of groups. We have suggested in Blaha (2021b), (2021e) and (2021g) that the set of hypercomplex space-time dimensions be transformed to a set symmetry groups in each HyperCosmos space.

There is a two-fold justification for this procedure: 1). There is no evidence for the existence of hypercomplex numbers in our universe. 2) The generated set of symmetry groups has the important purpose of providing ultra-weak interactions uniting Normal and Dark matter. Without unification, Dark matter becomes physically irrelevant fot elementary particle physics.

We begin by noting the space-time dimension is set by the dimension array:

$$r = \log_2 (d_{dN}/16) \qquad (11\text{-B.1})$$

For N = 9 the space-time dimension is 0 since d_{dN} = 16. For N = 7 (our universe's space) the space-time dimension is 4 since d_{dN} = 256. We will consider the cases of QUeST, UTMOST, and Maxiverse to illustrate the method.

11-B.1 Hypercomplex Coordinates Transformed to Symmetry Groups in Our Universe N = 7

Our QUeST formulation for our N = 7 universe has a d_{d7} = 256 component dimension array. A preliminary view of the symmetry groups that it implies appears in Fig. 11-B.1. Note that there are 32 dimensions that are initially allocated to space-time dimensions suggesting a hypercomplex space-time. We chose to allocate 4 real dimensions to obtain the 4-dimension space-time implied by eq. 11-B.1. The remaining 28 real dimensions we allocated to additional symmetry groups—namely seven U(2) groups. We call these groups Connection groups since their role is to "connect" fermions residing in different fermion spectrum blocks.

The structure of the seven additional U(2) groups is not specified. We chose to use a reasonable physical principle to allocate them. We believe their role is to "connect" fermions in different blocks since the fermions within each fermion block have "known" interactions. Note that there are initially eight blocks, each with their own set of symmetry groups and corresponding interactions, and initially no interactions between the eight blocks. If there are no block interactions (except gravity), then the Physics of the fermion set is conceptually disjoint. Thus we choose to implement inter-block interactions as in Fig. 11-B.2 following the principle:

A fermion in any block has interactions either directly, or indirectly, with every other fermion in any other block.

Fig. 11-B.2 shows an implementation of this principle. The horizontal lines in Fig. 11-B.2 indicate 1:1 transformations between all corresponding fermions of each "Normal" and "Dark" block. The three "angled" lines indicate 1:1 transformations between corresponding fermions of a "Normal" and a "Dark" fermion block in the layer above it. The result is a see-saw type of pattern.

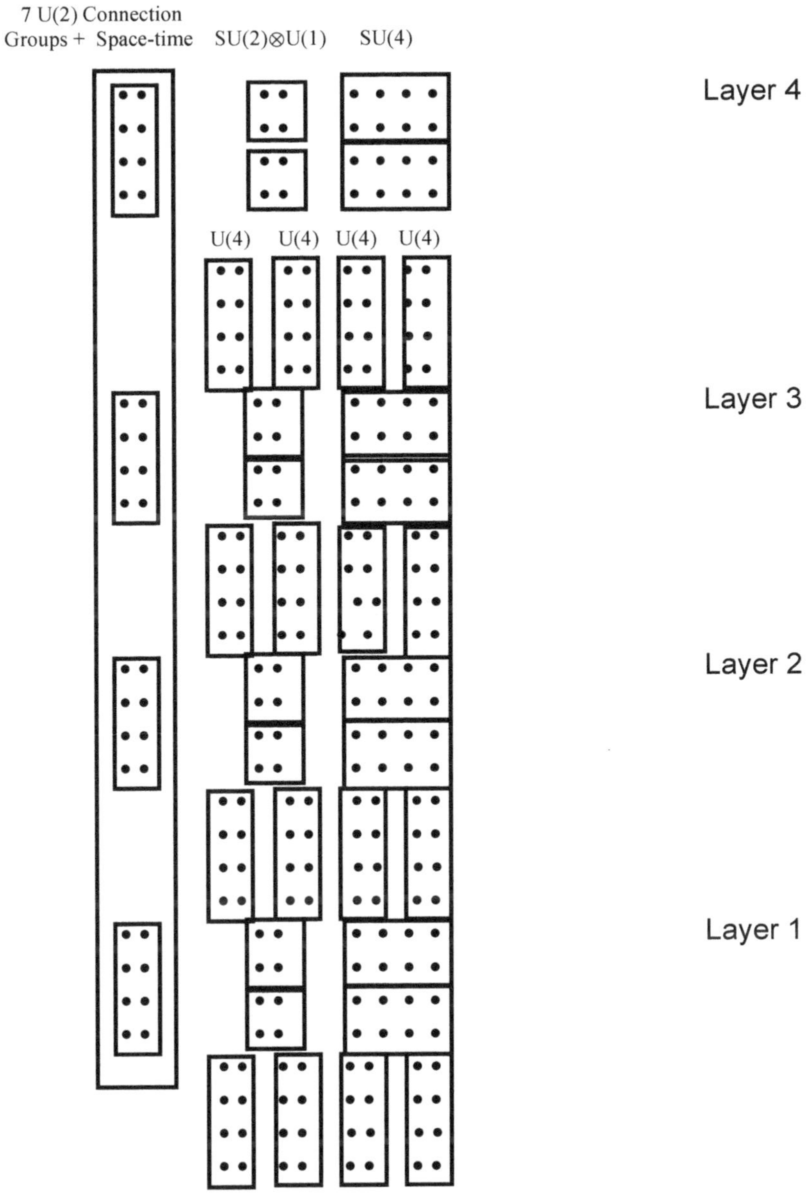

Figure 11-B.1. The four layers of QUeST, UST and QUeST internal symmetry groups (and space-time) with SU(4) before breakdown to SU(3)⊗U(1). Note the left column of blocks are combined below to specify a 4 dimension real space-time plus seven U(2) Connection groups. Note each layer has 64 dimensions = 56 + 8 dimensions.

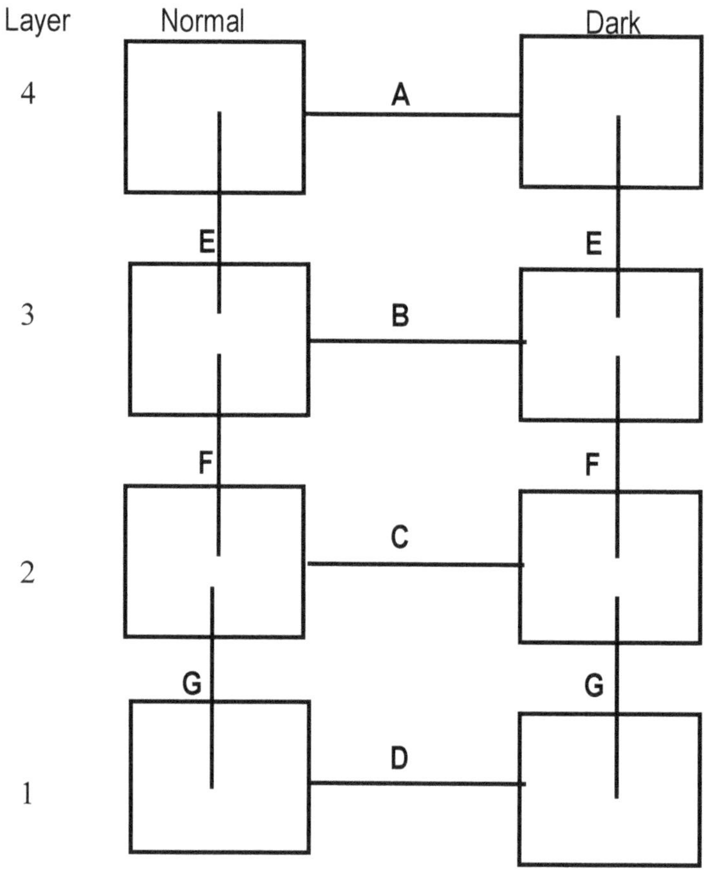

Figure 11-B.2. The seven U(2) Connection groups (shown as 10 lines) between the eight QUeST/UST blocks. Connection groups are obtained by transfering 28 dimensions from QUeST space-time to internal symmetries with the consequent reduction of the space-time from four octonion (complex quaternion) coordinates to four real coordinates. The Connection groups generate rotations and interactions between corresponding fermions and vector bosons of each pair of blocks. The Normal and Dark sector U(2) vertical connections above (E, F, G) represent the same U(2) groups.

11-B.1.1 The U(2) Connection Groups

The seven U(2) Connection groups of Fig. 11-B.2 generate "rotations" and interactions between *corresponding* fermions and vector bosons of each pair of blocks of the eight blocks of fermions in QUeST/UST.

11-B.1.1.1 Horizontal Lines

The horizontal lines in Fig. 11-B.2 (A, B, C, and D) each represent a U(2) Connection group that "rotates" two *corresponding* fermions in the Normal and Dark sectors of each layer. Thus a Normal e is "rotated" with a corresponding Dark e, and so on.

Each of the four horizontal Connection Groups has a reducible U(2) representation D that is the sum[50] of 4*8 = 32 irreducible U(2) representations. We may view the irreducible representations D_j as strung along the diagonal.

$$D = \sum_{j=1}^{32} D_j \qquad (11\text{-B.2})$$

for each of the U(2) groups of the four horizontal lines in Fig. 11-B.2.

The U(2) group also specifies gauge field interactions between corresponding fermions in each layer of the Normal and Dark sectors of the form

$$g\overline{\Psi}_{Nn}\gamma\cdot A\cdot T\Psi_{Dn} \qquad (11\text{-B.3})$$

where N indicates a Normal fermion and D indicates the corresponding Dark fermion, with A being a U(2) gauge vector boson, and n the label for corresponding fermions.

These U(2) transformations imply that the Normal and Dark sectors have the same species.

11-B.1.1.2 Vertical Lines

The pairs of vertical lines in Fig. 11-B.2 (E, F, G) each represent a U(2) Connection group that "rotates" sets of two *corresponding* fermions in adjacent layers as shown in Fig. 11-B.2 in the Normal and Dark sectors. Thus a Normal e in layer 1 is rotated with a corresponding Normal e in layer 2, and so on.

Each of the three (six counting both Normal and Dark lines in Fig. 11-B.2) vertical Connection Groups has a reducible U(2) representation D that is the sum of 64 irreducible U(2) representations.[51] We may view D as an array of 64 U(2) irreducible representation dimensions D_j strung along the diagonal.

$$D = \sum_{j=1}^{64} D_j \qquad (11\text{-B.4})$$

for each of the U(2) groups of the 3 (6) horizontal lines in Fig. 11-B.2. Note the 64 irreducible representations include both Normal and Dark sectors of a layer. [52]

Each U(2) group also specifies a gauge field interaction between corresponding fermions in adjacent layers for both Normal and Dark sectors:

---

[50] Eight fermions per generation ▪ four generations, thus accounting for each fermion in a block.
[51] Eight fermions per generations ▪ four generations ▪ 2 types of matter (Normal and Dark).
[52] There are 64 fermions in total for each of the four layers of NEWQUeST/NEWEST.

$$g\overline{\Psi}_{nl_1}\gamma\cdot A\cdot T\Psi_{nl_2} \qquad (11\text{-B}.5)$$

where l_1 and l_2 designate layers, A is a gauge field vector boson, and n the label for corresponding fermions.

Each E, F, and G U(2) group reducible representation includes both Normal and Dark sectors.

11-B.1.2 The Connection Groups are UltraWeak Interactions

Since there is no convincing experimental evidence for particle interactions between Normal and Dark matter, or between Normal fermion layers the Connection groups appear to be Ultraweak.

11-B.2 UTMOST with Six Real Space-Time Coordinates (Dimensions)

The UTMOST space is a Blaha number N = 6 HyperCosmos space. It is the Megaverse (Multiverse) with 6 real-valued space-time coordinates by eq.11-B.1 above. It has a d_{d6} = 1024 dimension array.

We view the UTMOST dimension array as composed of four copies of the N = 7 dimension array. Initially 4·32 = 128 dimensions are for space-time coordinates.

We allocate 4·28 = 112 dimensions to each of the four QUeST copies within UTMOST to give each copy 7 SU(2) Connection groups. Therefore 4* 28 = 112 of the 128 space-time dimensions of UTMOST are mapped. The remaining 16 dimensions give 6 real-valued space-time dimensions to the Megaverse space, 8 dimensions to a Megaverse SU(4) Connection group, and 2 dimensions to a U(1) group for all fermions in the Megaverse. See Figs. 11-B.3 and 11-B.4.

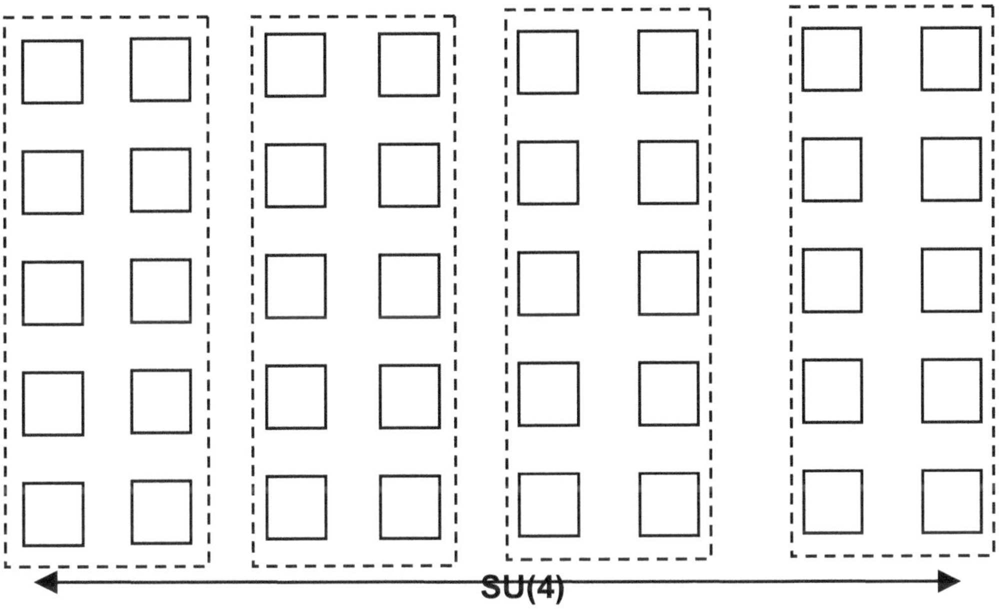

Figure 11-B.3. UTMOST has four QUeST copies. An SU(4) internal symmetry Connection group maps between corresponding fermions in the four copies: fermion by fermion. An additional U(1) Connection group applies to every corresponding fermion. It is not shown in this figure.

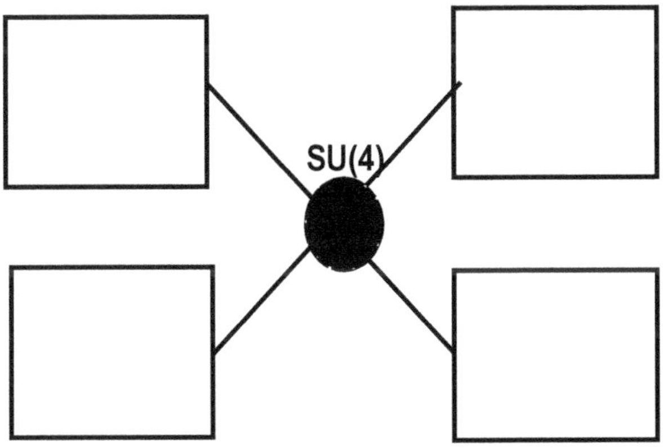

Figure 11-B.4. The SU(4) Connection Group of UTMOST connecting fermions in the four QUeST "copies" blocks. An additional U(1) Connection group applies to every corresponding fermion. It is not shown in this figure.

11-B.3 Maxiverse with Eight Real Coordinates (Dimensions)

The Maxiverse is a Blaha number N = 5 space with a dimension array of d_{d5} = 4096 dimensions.

The Maxiverse contains four copies of UTMOST. Each UTMOST has 4*6 = 24 space-time dimensions. (The remainder in each UTMOST copy consists of internal symmetries and Connection groups.) We allocate the 24 dimensions to eight real space-time dimensions plus 16 dimensions for a new ultraweak (possibly broken) SU(8) Connection group for the four parts of the Maxiverse. See Fig. 11-B.5.

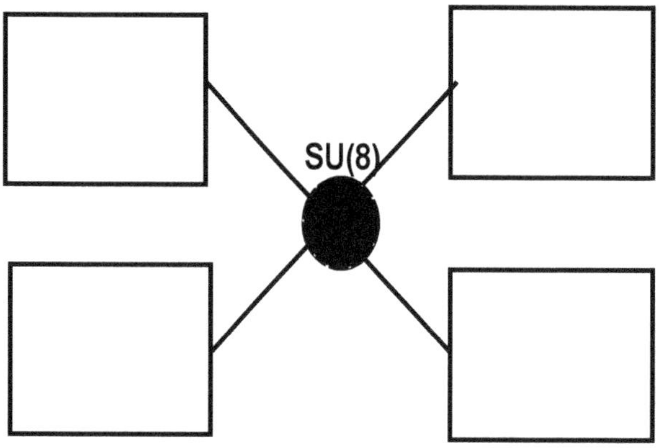

Figure 11-B.5. The SU(8) Connection Group of Maxiverse connecting fermions in the four UTMOST "copies" blocks.

11-B.4 Determining the Connection Groups for a Space

In the previous sections we determined the Connection Groups for N = 7, N = 6, and N = 5 spaces. The dimensions required for the Connection Groups of a space d_{gN} for N ≤ 6 can be obtained by noting the space-time dimension of a space-time r is related to the space-time dimension r – 2 of the next lower space. Since a space is four copies of the next lower space it has 4(r – 2) space-time dimensions from the four duplicates. Then, using eq. 2.17, we find

$$4r_{N+1} = 4(r_N - 2) = r_N + d_{gN}$$

or

$$d_{gN} = 3r_N - 8 \qquad (11\text{-B.6})$$

where d_{gN} is the *real-valued* space-time dimensions reallocated to Connection Groups in space N. For N = 6, r_6 = 6, d_{gN} = 10. For N = 5, r_5 = 8, d_{gN} = 16. And so on.

Eq. 11-B.6 determines the dimensions available for Connection Groups for the space with Blaha number N ≤ 6. Using eq. 11-B.1:

$$r_N = \log_2 (d_{dN}/16) \qquad (11\text{-B.7})$$

we relate d_{gN} to the size of the dimension array:

$$d_{gN} = 3 \log_2(d_{dN}/16) - 8 \qquad (11\text{-B.8})$$

thus specifying the number of space-time dimensions allocated to space number N.

REFERENCES

Akhiezer, N. I., Frink, A. H. (tr), 1962, *The Calculus of Variations* (Blaisdell Publishing, New York, 1962).

Bjorken, J. D., Drell, S. D., 1964, *Relativistic Quantum Mechanics* (McGraw-Hill, New York, 1965).

Bjorken, J. D., Drell, S. D., 1965, *Relativistic Quantum Fields* (McGraw-Hill, New York, 1965).

Blaha, S., 1995, *C++ for Professional Programming* (International Thomson Publishing, Boston, 1995).

_______, 1998, *Cosmos and Consciousness* (Pingree-Hill Publishing, Auburn, NH, 1998 and 2002).

_______, 2002, *A Finite Unified Quantum Field Theory of the Elementary Particle Standard Model and Quantum Gravity Based on New Quantum Dimensions™ & a New Paradigm in the Calculus of Variations* (Pingree-Hill Publishing, Auburn, NH, 2002).

_______, 2004, *Quantum Big Bang Cosmology: Complex Space-time General Relativity, Quantum Coordinates™ Dodecahedral Universe, Inflation, and New Spin 0, ½, 1 & 2 Tachyons & Imagyons* (Pingree-Hill Publishing, Auburn, NH, 2004).

______, *2005a, Quantum Theory of the Third Kind: A New Type of Divergence-free Quantum Field Theory Supporting a Unified Standard Model of Elementary Particles and Quantum Gravity based on a New Method in the Calculus of Variations* (Pingree-Hill Publishing, Auburn, NH, 2005).

______, 2005b, *The Metatheory of Physics Theories, and the Theory of Everything as a Quantum Computer Language* (Pingree-Hill Publishing, Auburn, NH, 2005).

______, 2005c, *The Equivalence of Elementary Particle Theories and Computer Languages: Quantum Computers, Turing Machines, Standard Model, Superstring Theory, and a Proof that Gödel's Theorem Implies Nature Must Be Quantum* (Pingree-Hill Publishing, Auburn, NH, 2005).

______, 2006a, *The Foundation of the Forces of Nature* (Pingree-Hill Publishing, Auburn, NH, 2006).

______, 2006b, *A Derivation of ElectroWeak Theory based on an Extension of Special Relativity; Black Hole Tachyons; & Tachyons of Any Spin.* (Pingree-Hill Publishing, Auburn, NH, 2006).

_______, 2007a, *Physics Beyond the Light Barrier: The Source of Parity Violation, Tachyons, and A Derivation of Standard Model Features* (Pingree-Hill Publishing, Auburn, NH, 2007).

______, 2007b, *The Origin of the Standard Model: The Genesis of Four Quark and Lepton Species, Parity Violation, the ElectroWeak Sector, Color SU(3), Three Visible Generations of Fermions, and One Generation of Dark Matter with Dark Energy* (Pingree-Hill Publishing, Auburn, NH, 2007).

______, *2008a, A Direct Derivation of the Form of the Standard Model From GL(16) (Pingree-Hill Publishing, Auburn, NH, 2008).*

______, 2008b, *A Complete Derivation of the Form of the Standard Model With a New Method to Generate Particle Masses Second Edition* (Pingree-Hill Publishing, Auburn, NH, 2008)

______, 2009, *The Algebra of Thought & Reality: The Mathematical Basis for Plato's Theory of Ideas, and Reality Extended to Include A Priori Observers and Space-Time Second Edition* (Pingree-Hill Publishing, Auburn, NH, 2009).

______, 2010a, *Operator Metaphysics: A New Metaphysics Based on a New Operator Logic and a New Quantum Operator Logic that Lead to a Mathematical Basis for Plato's Theory of Ideas and Reality* (Pingree-Hill Publishing, Auburn, NH, 2010).

______, 2010b, *The Standard Model's Form Derived from Operator Logic, Superluminal Transformations and GL(16)* (Pingree-Hill Publishing, Auburn, NH, 2010).

______, 2010c, *SuperCivilizations: Civilizations as Superorganisms* (McMann-Fisher Publishing, Auburn, NH, 2010).

______, 2011a, *21st Century Natural Philosophy Of Ultimate Physical Reality* (McMann-Fisher Publishing, Auburn, NH, 2011).

______, 2011b, *All the Universe! Faster Than Light Tachyon Quark Starships & Particle Accelerators with the LHC as a Prototype Starship Drive Scientific Edition* (Pingree-Hill Publishing, Auburn, NH, 2011).

______, 2011c, *From Asynchronous Logic to The Standard Model to Superflight to the Stars* (Blaha Research, Auburn, NH, 2011).

______, 2012a, *From Asynchronous Logic to The Standard Model to Superflight to the Stars volume 2: Superluminal CP and CPT, U(4) Complex General Relativity and The Standard Model, Complex Vierbein General Relativity, Kinetic Theory, Thermodynamics* (Blaha Research, Auburn, NH, 2012).

______, 2012b, *Standard Model Symmetries, And Four And Sixteen Dimension Complex Relativity; The Origin Of Higgs Mass Terms* (Blaha Reasearch, Auburn, NH, 2012).

______, 2013a, *Multi-Stage Space Guns, Micro-Pulse Nuclear Rockets, and Faster-Than-Light Quark-Gluon Ion Drive Starships* (Blaha Research, Auburn, NH, 2013).

______, 2013b, *The Bridge to Dark Matter; A New Sibling Universe; Dark Energy; Inflatons; Quantum Big Bang; Superluminal Physics; An Extended Standard Model Based on Geometry* (Blaha Reasearch, Auburn, NH, 2013).

______, 2014a, *Universes and Megaverses: From a New Standard Model to a Physical Megaverse; The Big Bang; Our Sibling Universe's Wormhole; Origin of the Cosmological Constant, Spatial Asymmetry of the Universe, and its Web of Galaxies; A Baryonic Field between Universes and Particles; Megaverse Extended Wheeler-DeWitt Equation* (Blaha Reasearch, Auburn, NH, 2014).

______, 2014b, *All the Megaverse! Starships Exploring the Endless Universes of the Cosmos Using the Baryonic Force* (Blaha Research, Auburn, NH, 2014).

______, 2014c, *All the Megaverse! II Between Megaverse Universes: Quantum Entanglement Explained by the Megaverse Coherent Baryonic Radiation Devices – PHASERs Neutron Star Megaverse Slingshot Dynamics Spiritual and UFO Events, and the Megaverse Microscopic Entry into the Megaverse* (Blaha Research, Auburn, NH, 2014).

______, 2015a, *PHYSICS IS LOGIC PAINTED ON THE VOID: Origin of Bare Masses and The Standard Model in Logic, U(4) Origin of the Generations, Normal and Dark Baryonic Forces, Dark Matter, Dark Energy, The Big Bang, Complex General Relativity, A Megaverse of Universe Particles* (Blaha Research, Auburn, NH, 2015).

______, 2015b, *PHYSICS IS LOGIC Part II: The Theory of Everything, The Megaverse Theory of Everything, U(4)⊗U(4) Grand Unified Theory (GUT), Inertial Mass = Gravitational Mass, Unified Extended Standard Model and a New Complex General Relativity with Higgs Particles, Generation Group Higgs Particles* (Blaha Research, Auburn, NH, 2015).

______, 2015c, *The Origin of Higgs ("God") Particles and the Higgs Mechanism: Physics is Logic III, Beyond Higgs – A Revamped Theory With a Local Arrow of Time, The Theory of Everything Enhanced, Why Inertial Frames are Special, Universes of the Mind* (Blaha Research, Auburn, NH, 2015).

______, 2015d, *The Origin of the Eight Coupling Constants of The Theory of Everything: U(8) Grand Unified Theory of Everything (GUTE), S^8 Coupling Constant Symmetry, Space-Time Dependent Coupling Constants, Big Bang Vacuum Coupling Constants, Physics is Logic IV* (Blaha Research, Auburn, NH, 2015).

______, 2016a, *New Types of Dark Matter, Big Bang Equipartition, and A New U(4) Symmetry in the Theory of Everything: Equipartition Principle for Fermions, Matter is 83.33% Dark, Penetrating the Veil of the Big Bang, Explicit QFT Quark Confinement and Charmonium, Physics is Logic V* (Blaha Research, Auburn, NH, 2016).

______, 2016b, *The Periodic Table of the 192 Quarks and Leptons in The Theory of Everything: The U(4) Layer Group, Physics is Logic VI* (Blaha Research, Auburn, NH, 2016).

______, 2016c, *New Boson Quantum Field Theory, Dark Matter Dynamics, Dark Matter Fermion Layer Mixing, Genesis of Higgs Particles, New Layer Higgs Masses, Higgs Coupling Constants, Non-Abelian Higgs Gauge Fields, Physics is Logic VII* (Blaha Research, Auburn, NH, 2016).

______, 2016d, *Unification of the Strong Interactions and Gravitation: Quark Confinement Linked to Modified Short-Distance Gravity; Physics is Logic VIII* (Blaha Research, Auburn, NH, 2016).

______, 2016e, *MoND: Unification of the Strong Interactions and Gravitation II, Quark Confinement Linked to Large-Scale Gravity, Physics is Logic IX* (Blaha Research, Auburn, NH, 2016).

______, 2016f, *CQ Mechanics: A Unification of Quantum & Classical Mechanics, Quantum/Semi-Classical Entanglement, Quantum/Classical Path Integrals, Quantum/Classical Chaos* (Blaha Research, Auburn, NH, 2016).

______, 2016g, *GEMS Unified Gravity, ElectroMagnetic and Strong Interactions: Manifest Quark Confinement, A Solution for the Proton Spin Puzzle, Modified Gravity on the Galactic Scale* (Pingree Hill Publishing, Auburn, NH, 2016).

______, 2016h, *Unification of the Seven Boson Interactions based on the Riemann-Christoffel Curvature Tensor* (Pingree Hill Publishing, Auburn, NH, 2016).

______, 2017a, *Unification of the Eleven Boson Interactions based on 'Rotations of Interactions'* (Pingree Hill Publishing, Auburn, NH, 2017).

______, 2017b, *The Origin of Fermions and Bosons, and Their Unification* (Pingree Hill Publishing, Auburn, NH, 2017).

______, 2017c, *Megaverse: The Universe of Universes* (Pingree Hill Publishing, Auburn, NH, 2017).

______, 2017d, *SuperSymmetry and the Unified SuperStandard Model* (Pingree Hill Publishing, Auburn, NH, 2017).

______, 2017e, *From Qubits to the Unified SuperStandard Model with Embedded SuperStrings: A Derivation* (Pingree Hill Publishing, Auburn, NH, 2017).

______, 2017f, *The Unified SuperStandard Model in Our Universe and the Megaverse: Quarks, ... ,* (Pingree Hill Publishing, Auburn, NH, 2017).

______, 2018a, *The Unified SuperStandard Model and the Megaverse SECOND EDITION A Deeper Theory based on a New Particle Functional Space that Explicates Quantum Entanglement Spookiness (Volume 1)* (Pingree Hill Publishing, Auburn, NH, 2018).

______, 2018b, *Cosmos Creation: The Unified SuperStandard Model, Volume 2, SECOND EDITION* (Pingree Hill Publishing, Auburn, NH, 2018).

______, 2018c, *God Theory (*Pingree Hill Publishing, Auburn, NH, 2018).

______, 2018d, *Immortal Eye: God Theory: Second Edition* (Pingree Hill Publishing, Auburn, NH, 2018).

______, 2018e, *Unification of God Theory and Unified SuperStandard Model THIRD EDITION* (Pingree Hill Publishing, Auburn, NH, 2018).

______, 2019a, *Calculation of: QED α = 1/137, and Other Coupling Constants of the Unified SuperStandard Theory* (Pingree Hill Publishing, Auburn, NH, 2019).

______, 2019b, *Coupling Constants of the Unified SuperStandard Theory SECOND EDITION* (Pingree Hill Publishing, Auburn, NH, 2019).

______, 2019c, *New Hybrid Quantum Big_Bang–Megaverse_Driven Universe with a Finite Big Bang and an Increasing Hubble Constant* (Pingree Hill Publishing, Auburn, NH, 2019).

______, 2019d, *The Universe, The Electron and The Vacuum* (Pingree Hill Publishing, Auburn, NH, 2019).

______, 2019e, *Quantum Big Bang – Quantum Vacuum Universes (Particles)* (Pingree Hill Publishing, Auburn, NH, 2019).

______, 2019f, *The Exact QED Calculation of the Fine Structure Constant Implies ALL 4D Universes have the Same Physics/Life Prospects* (Pingree Hill Publishing, Auburn, NH, 2019).

______, 2019g, *Unified SuperStandard Theory and the SuperUniverse Model: The Foundation of Science* (Pingree Hill Publishing, Auburn, NH, 2019).

______, 2020a, *Quaternion Unified SuperStandard Theory (The QUeST) and Megaverse Octonion SuperStandard Theory (MOST)* (Pingree Hill Publishing, Auburn, NH, 2020).

______, 2020b, *United Universes Quaternion Universe - Octonion Megaverse* (Pingree Hill Publishing, Auburn, NH, 2020).

______, 2020c, *Unified SuperStandard Theories for Quaternion Universes & The Octonion Megaverse* (Pingree Hill Publishing, Auburn, NH, 2020).

______, 2020d, *The Essence of Eternity: Quaternion & Octonion SuperStandard Theories* (Pingree Hill Publishing, Auburn, NH, 2020).

______, 2020e, *The Essence of Eternity II* (Pingree Hill Publishing, Auburn, NH, 2020).

______, 2020f, *A Very Conscious Universe* (Pingree Hill Publishing, Auburn, NH, 2020).

______, 2020g, *Hypercomplex Universe* (Pingree Hill Publishing, Auburn, NH, 2020).

______, 2020h, *Beneath the Quaternion Universe* (Pingree Hill Publishing, Auburn, NH, 2020).

______, 2020i, *Why is the Universe Real? From Quaternion & Octonion to Real Coordinates* (Pingree Hill Publishing, Auburn, NH, 2020).

______, 2020j, *The Origin of Universes: of Quaternion Unified SuperStandard Theory (QUeST); and of the Octonion Megaverse (UTMOST)* (Pingree Hill Publishing, Auburn, NH, 2020).

______, 2020k, *The Seven Spaces of Creation: Octonion Cosmology* (Pingree Hill Publishing, Auburn, NH, 2020).

______, 2020l, *From Octonion Cosmology to the Unified SuperStandard Theory of Particles* (Pingree Hill Publishing, Auburn, NH, 2020).

______, 2021a, *Pioneering the Cosmos* (Pingree Hill Publishing, Auburn, NH, 2021).

______, 2021b, *Pioneering the Cosmos II* (Pingree Hill Publishing, Auburn, NH, 2021).

______, 2021c, *Beyond Octonion Cosmology* (Pingree Hill Publishing, Auburn, NH, 2021).

______, 2021d, *Universes are Particles* (Pingree Hill Publishing, Auburn, NH, 2021).

______, 2021e, *Octonion-like dna-based life, Universe expansion is decay, Emerging New Physics* (Pingree Hill Publishing, Auburn, NH, 2021).

______, 2021f, *The Science of Creation New Quantum Field Theory of Spaces* (Pingree Hill Publishing, Auburn, NH, 2021).

______, 2021g, *Quantum Space Theory With Application to Octonion Cosmology & Possibly To Fermionic Condensed Matter* (Pingree Hill Publishing, Auburn, NH, 2021).

______, 2021h, *21st Century Natural Philosophy of Octonion Cosmology , and Predestination, Fate, and Free Will* (Pingree Hill Publishing, Auburn, NH, 2021).

______, 2021i, *Beyond Octonion Cosmology II : Origin of the Quantum; A New Generalized Field Theory (GiFT); A Proof of the Spectrum of Universes; Atoms in Higher Universes* (Pingree Hill Publishing, Auburn, NH, 2021).

______, 2021j, *Integration of General Relativity and Quantum Theory: Octonion Cosmology, GiFT, Creation/Annihilation Spaces CASe, Reduction of Spaces to a Few Fermions and Symmetries in Fundamental Frames* (Pingree Hill Publishing, Auburn, NH, 2021).

______, 2022a, *New View of Octonion Cosmology Based on the Unification of General Relativit and Quantum Theory* (Pingree Hill Publishing, Auburn, NH, 2022).

______, 2022b, *The Gold Dust Beneath Hypercomplex Cosmology* (Pingree Hill Publishing, Auburn, NH, 2022).

______, 2022c, *Passing Through Nature to Eternity: ProtoCosmos, HyperCosmos, Unified SuperStandard Theory* (Pingree Hill Publishing, Auburn, NH, 2022).

______, 2022d, *HyperCosmos Fractionation and Fundamental Reference Frame Based Unification: Particle Inner Space Basis of Parton and Dual Resonance Models* (Pingree Hill Publishing, Auburn, NH, 2022).

______, 2022e, *A New UniDimension ProtoCosmos and SuperString F-Theory Relation to the HyperCosmos* (Pingree Hill Publishing, Auburn, NH, 2022).

______, 2022f, *The Cosmic Panorama: ProtoCosmos, HyperCosmos, Unified SuperStandard Theory (UST) Derivation* (Pingree Hill Publishing, Auburn, NH, 2022).

______, 2022g, *Ultimate Origin: ProtoCosmos and HyperCosmos* (Pingree Hill Publishing, Auburn, NH, 2022).

Eddington, A. S., 1952, *The Mathematical Theory of Relativity* (Cambridge University Press, Cambridge, U.K., 1952).

Fant, Karl M., 2005, *Logically Determined Design: Clockless System Design With NULL Convention Logic* (John Wiley and Sons, Hoboken, NJ, 2005).

Feinberg, G. and Shapiro, R., 1980, *Life Beyond Earth: The Intelligent Earthlings Guide to Life in the Universe* (William Morrow and Company, New York, 1980).

Gelfand, I. M., Fomin, S. V., Silverman, R. A. (tr), 2000, *Calculus of Variations* (Dover Publications, Mineola, NY, 2000).

Giaquinta, M., Modica, G., Souchek, J., 1998, *Cartesian Coordinates in the Calculus of Variations* Volumes I and II (Springer-Verlag, New York, 1998).

Giaquinta, M., Hildebrandt, S., 1996, *Calculus of Variations* Volumes I and II (Springer-Verlag, New York, 1996).

Gradshteyn, I. S. and Ryzhik, I. M., 1965, *Table of Integrals, Series, and Products* (Academic Press, New York, 1965).

Heitler, W., 1954, *The Quantum Theory of Radiation* (Claendon Press, Oxford, UK, 1954).

Huang, Kerson, 1992, *Quarks, Leptons & Gauge Fields 2nd Edition* (World Scientific Publishing Company, Singapore, 1992).

Jost, J., Li-Jost, X., 1998, *Calculus of Variations* (Cambridge University Press, New York, 1998).

Kaku, Michio, 1993, *Quantum Field Theory*, (Oxford University Press, New York, 1993).

Kirk, G. S. and Raven, J. E., 1962, *The Presocratic Philosophers* (Cambridge University Press, New York, 1962).

Landau, L. D. and Lifshitz, E. M., 1987, *Fluid Mechanics 2nd Edition*, (Pergamon Press, Elmsford, NY, 1987).

Misner, C. W., Thorne, K. S., and Wheeler, J. A., 1973, *Gravitation* (W. H. Freeman, New York, 1973).

Rescher, N., 1967, *The Philosophy of Leibniz* (Prentice-Hall, Englewood Cliffs, NJ, 1967).

Rieffel, Eleanor and Polak, Wolfgang, 2014, *Quantum Computing* (MIT Press, Cambridge, MA, 2014).

Riesz, Frigyes and Sz.-Nagy, Béla, 1990, *Functional Analysis* (Dover Publications, New York, 1990).

Sagan, H., 1993, *Introduction to the Calculus of Variations* (Dover Publications, Mineola, NY, 1993).

Sakurai, J. J., 1964, *Invariance Principles and Elementary Particles* (Princeton University Press, Princeton, NJ, 1964).

Weinberg, S., 1972, *Gravitation and Cosmology* (John Wiley and Sons, New York, 1972).

Weinberg, S., 1995, *The Quantum Theory of Fields Volume I* (Cambridge University Press, New York, 1995).

INDEX

The Author's Explorations

Stephen Blaha is a well-known Physicist and Man of Letters with interests in Science, Society and civilization, the Arts, and Technology. He had an Alfred P. Sloan Foundation scholarship in college. He received his Ph.D. in Physics from Rockefeller University. He has served on the faculties of several major universities. He was also a Member of the Technical Staff at Bell Laboratories, a manager at the Boston Globe Newspaper, a Director at Wang Laboratories, and President of Blaha Software Inc. and of Janus Associates Inc. (NH).

Among other achievements he was a co-discoverer of the "r potential" for heavy quark binding developing the first (and still the only demonstrable) non-Aeolian gauge theory with an "r" potential; first suggested the existence of topological structures in superfluid He-3; first proposed Yang-Mills theories would appear in condensed matter phenomena with non-scalar order parameters; first developed a grammar-based formalism for quantum computers and applied it to elementary particle theories; first developed a new form of quantum field theory without divergences (thus solving a major 60 year old problem that enabled a unified theory of the Standard Model and Quantum Gravity without divergences to be developed); first developed a formulation of complex General Relativity based on analytic continuation from real space-time; first developed a generalized non-homogeneous Robertson-Walker metric that enabled a quantum theory of the Big Bang to be developed without singularities at t = 0; first generalized Cauchy's theorem and Gauss' theorem to complex, curved multi-dimensional spaces; received Honorable Mention in the Gravity Research Foundation Essay Competition in 1978; first developed a physically acceptable theory of faster-than-light particles; first derived a composition of extremums method in the Calculus of Variations; first quantitatively suggested that inflationary periods in the history of the universe were not needed; first proved Gödel's Theorem implies Nature must be quantum; provided a new alternative to the Higgs Mechanism, and Higgs particles, to generate masses; first showed how to resolve logical paradoxes including Gödel's Undecidability Theorem by developing Operator Logic and Quantum Operator Logic; first developed a quantitative harmonic oscillator-like model of the life cycle, and interactions, of civilizations; first showed how equations describing superorganisms also apply to civilizations. A recent book shows his theory applies successfully to the past 14 years of history and to *new* archaeological data on Andean and Mayan civilizations as well as Early Anatolian and Egyptian civilizations.

He first developed an axiomatic derivation of the form of The Standard Model from geometry – space-time properties – The Unified SuperStandard Model. It unifies all the known forces of Nature. It also has a Dark Matter sector that includes a Dark ElectroWeak sector with Dark doublets and Dark gauge interactions. It uses quantum coordinates to remove infinities that crop up in most

interacting quantum field theories and additionally to remove the infinities that appear in the Big Bang and generate inflationary growth of the universe. It shows gravity has a MOND-like form without sacrificing Newton's Laws. It relates the interactions of the MOND-like sector of gravity with the r-potential of Quark Confinement. The axioms of the theory lead to the question of their origin. We suggest in the preceding edition of this book it can be attributed to an entity with God-like properties. We explore these properties in "God Theory" and show they predict that the Cosmos exists forever although individual universes (or incarnations of our universe) "come and go." Several other important results emerge from God Theory such a functionally triune God. The Unified SuperStandard Theory has many other important parts described in the Current Edition of *The Unified SuperStandard Theory* and expanded in subsequent volumes.

Blaha has had a major impact on a succession of elementary particle theories: his Ph.D. thesis (1970), and papers, showed that quantum field theory calculations to all orders in ladder approximations could not give scaling deep inelastic electron-nucleon scattering. He later showed the eigenvalue equation for the fine structure constant α in Johnson-Baker-Willey QED had a zero at $\alpha = 1$ not 1/137 by solving the Schwinger-Dyson equations to all orders in an approximation that agreed with exact results to 4th order in α thus ending interest in this theory. In 1979 at Prof. Ken Johnson's (MIT) suggestion he calculated the proton-neutron mass difference in the MIT bag model and found the result had the wrong sign reducing interest in the bag model. These results all appear in Physical Review papers. In the 2000's he repeatedly pointed out the shortcomings of SuperString theory and showed that The Standard Model's form could be derived from space-time geometry by an extension of Lorentz transformations to faster than light transformations. This deeper space-time basis greatly increases the possibility that it is part of THE fundamental theory. Recently, Blaha showed that the Weak interactions differed significantly from the Strong, electromagnetic and gravitation interactions in important respects while these interactions had similar features, and suggested that ElectroWeak theory, which is essentially a glued union of the Weak interactions and Electromagnetism, possibly modulo unknown Higgs particle features, be replaced by a unified theory of the other interactions combined with a stand-alone Weak interaction theory. Blaha also showed that, if Charmonium calculations are taken seriously, the Strong interaction coupling constant is only a factor of five larger than the electromagnetic coupling constant, and thus Strong interaction perturbation theory would make sense and yield physically meaningful results.

In graduate school (1965-71) he wrote substantial papers in elementary particles and group theory: The Inelastic E- P Structure Functions in a Gluon Model. Phys. Lett. B40:501-502,1972; Deep-Inelastic E-P Structure Functions In A Ladder Model With Spin 1/2 Nucleons, Phys.Rev. D3:510-523,1971; Continuum Contributions To The Pion Radius, Phys. Rev. 178:2167-2169,1969; Character Analysis of U(N) and SU(N), J. Math. Phys. 10, 2156 (1969); and The Calculation of the Irreducible Characters of the Symmetric Group in Terms of the

Compound Characters, (Published as Blaha's Lemma in D. E. Knuth's book: *The Art of Computer Programming Vols. 1 – 4*).

In the early 1980's Blaha was also a pioneer in the development of UNIX for financial, scientific and Internet applications: benchmarked UNIX versions showing that block size was critical for UNIX performance, developing financial modeling software, starting database benchmarking comparison studies, developing Internet-like UNIX networking (1982) and developing a hybrid shell programming technique (1982) that was a precursor to the PERL programming language. He was also the manager of the AT&T ten-year future products development database. His work helped lead to commercial UNIX on computers such as Sun Micros, IBM AIX minis, and Apple computers.

In the 1980's he pioneered the development of PC Desktop Publishing on laser printers and was nominated for three "Awards for Technical Excellence" in 1987 by PC Magazine for PC software products that he designed and developed.

Recently he has developed a theory of Megaverses – actual universes of which our universe is one – with quantum particle-like properties based on the Wheeler-DeWitt equation of Quantum Gravity. He has developed a theory of a baryonic force, which had been conjectured many years ago, and estimated the strength of the force based on discrepancies in measurements of the gravitational constant G. This force, operative in D-dimensional space, can be used to escape from our universe in "uniships" which are the equivalent of the faster-than-light starships proposed in the author's earlier books. Thus travel to other universes, as well as to other stars is possible.

Blaha also considered the complexified Wheeler-DeWitt equation and showed that its limitation to real-valued coordinates and metrics generated a Cosmological Constant in the Einstein equations.

The author has also recently written a series of books on the serious problems of the United States and their solution as well as a book on the decline of Mankind that will follow from current social and genetic trends in Mankind.

In the past twenty years Dr. Blaha has written over 80 books on a wide range of topics. Some recent major works are: *From Asynchronous Logic to The Standard Model to Superflight to the Stars*, *All the Universe!*, *SuperCivilizations: Civilizations as Superorganisms, America's Future: an Islamic Surge, ISIS, al Qaeda, World Epidemics, Ukraine, Russia-China Pact, US Leadership Crisis, The Rises and Falls of Man – Destiny – 3000 AD: New Support for a Superorganism MACRO-THEORY of CIVILIZATIONS From CURRENT WORLD TRENDS and NEW Peruvian, Pre-Mayan, Mayan, Anatolian, and Early Egyptian Data, with a Projection to 3000 AD,* and *Mankind in Decline: Genetic Disasters, Human-Animal Hybrids, Overpopulation, Pollution, Global Warming, Food and Water Shortages, Desertification, Poverty, Rising Violence, Genocide, Epidemics, Wars, Leadership Failure.*

He has taught approximately 4,000 students in undergraduate, graduate, and postgraduate corporate education courses primarily in major universities, and large companies and government agencies.

He developed a quantum theory, The Unified SuperStandard Theory (UST), which describes elementary particles in detail without the difficulties of conventional quantum field theory. He found that the internal symmetries of this theory could be exactly derived from an octonion theory called QUeST. He further found that another octonion theory (UTMOST) describes the Megaverse. It can hold QUeST universes such as our own universe. It has an internal symmetry structure which is a superset of the QUeST internal symmetries.

Recently he developed Octonion Cosmology. He replaced it with HyperCosmos theory, which has significantly better features. He developed a fractionalization process for dimensions, particles and symmetry groups. He also described transformation that reduced particles and dimensions to a far more compact form. He also developed a precursor theory ProtoCosmos that leads to the HyperCosmos.

The author showed that space-time and Internal Symmetries can be unified in any of the ten HyperCosmos spaces in their associated HyperUnification spaces. The combined set of HyperUnification spaces enable all HyperCosmos dimensions to be obtained by a General Relativistic transformation from one primordial dimension in the 42 space-time dimension unified HyperUnification space.

www.ingramcontent.com/pod-product-compliance
Ingram Content Group UK Ltd.
Pitfield, Milton Keynes, MK11 3LW, UK
UKHW060220061026
15625UKWH00007B/408

9 798985 826661